Palestine - Peace by Piece

Ahmed Abukhater

Palestine - Peace by Piece

Transformative Conflict Resolution for Land and Trans-boundary Water Resources

Springer

Ahmed Abukhater
Denver, CO, USA

ISBN 978-3-030-03984-4 ISBN 978-3-030-03985-1 (eBook)
https://doi.org/10.1007/978-3-030-03985-1

Library of Congress Control Number: 2018964229

This Springer imprint is published by the registered company Springer Nature Switzerland AG
The registered company address is: Gewerbestrasse 11, 6330 Cham, Switzerland

This book is dedicated to five very special people...

...my beloved parents Baha' El-Deen *and* Fatima, *who showed me the real meaning of unconditional love and taught me that with faith and determination you can overcome any obstacle no matter how long or difficult the journey might be ... that justice will prevail no matter how long injustice endures ... and that the darkness of night can only last for so long before an astounding sunrise breaks through on the horizon declaring a new day.*

...the departed *whose footprints charted the way to victory and freedom.*

...my only love, *you will always be in my heart.*

...you, the reader*; your human nature and free spirit.*

Preface

As I begin to write this book, I am overcome with emotions and the memories of a time never to return—growing up in Palestine, listening to my late grandmother's bedtime stories about life before the existence of Israel in a land whose inhabitants loved it so dearly. It feels as though I have written this story a thousand times but nothing can truly explain what this really means to me as a Palestinian refugee whose family endured the horror of displacement and unimaginable persecution, the agony of forced expulsion from one's own homeland, and the ongoing daily hardship and misery of living for decades under grinding occupation. Our story is the story of all Palestinians inside the occupied territories and in exile. It is the extraordinary story of those who still endure the horror of living in refugee camps in the Gaza Strip, in what has become the world's largest open-air prison, in the shadow of one of the most brutal military and economic sieges in human history, waiting for a resolution that never materialized. It is a story that has yet to find an end and that is so desperately looking for a resolution; a story worth telling in every aspect.

This book, however, is not intended to be a personal memoir or to cover specific historical accounts, nor is it intended to demonize one side or the other. Rather, this book will objectively narrate key historical events that shaped the current situation in hopes for achieving lasting peace and security for all parties involved. My goal is to shed light on the untold side of the story, the one that is well known and understood by many who witnessed and lived through these events, with great emphasis on the basic and universal human values that bind us all together and make us strong with a pronounced sense of humanity. In this book, I will utilize my personal story and experience, which I seek to integrate into a much larger and more meaningful context for the sake of charting a course to finding peaceful closure to our longstanding suffering and persecution. By so doing, I seek to delineate, as dispassionately as I can, a transformative, operational framework and a pragmatic roadmap for peace to ensure sustainable peace and an end to the current conflict, and most importantly justice for all. By providing this framework, I hope to provide a tool that can be utilized to pave the way for a meaningful resolution. It is paramount to understand that this is only a framework and not a solution. This is due in part to the fact that the challenge that this intractable dispute presents is an adaptive one not technical in nature. As such, we cannot apply technical solutions to such an adaptive problem. There is no cut and dried solution for this conflict. It is only our ability to successfully and consistently implement this framework, coupled with the disciplined and thoughtful adherence to its guidelines that would determine how successful the resultant solution will be. Akin to other frameworks, this framework is not intended to be viewed as an exclusive solution, but rather an alternative way to the *status quo* "peace process," whose failure has become irrefutable. It is an ambitious, yet given the political will, attainable goal towards which this book hopes to contribute.

My story, which began with my childhood growing up in a small temporary United Nations shelter that eventually turned into our home, brings wonderful but also haunting memories of living with the harsh realities and cruelty of military occupation and the systematic oppression that was imposed on us, young and old. Throughout my whole life as a child and a young adult, I knew very little about what normal upbringing is actually like. Like many other Palestinians, I was living in a constant mode of survival and triage, in which one can only aspire to make it through the day, not knowing what tomorrow would bring, let alone planning for a meaningful future.

The first *Intifada* (uprising) in 1987 was a defining moment in my life. Going to school as a child, for example, was such an ordeal. The choice was clear; go to school and you may risk getting shot or stay home and you might be safe only for a while. I made my choice and so did everyone living in the camp. Living in the Brazil refugee camp in Rafah south of Gaza meant that we had to fight our way to go to school every day with death haunting us at every corner. Going to school was akin to going to a battlefield, trying to cross the street with the blinding smoke of tear gas and the smell of death everywhere. Running from one corner of the street to the other through tear gas, live bullets and ammunition was just part of the daily commute to school. I can still vividly remember the crisp smell of the morning breeze mixed with the constant and overwhelming odor of tear gas, and fear. As kids, we used to hold our breath just long enough to run across the main square to get to the other side which was so blackened with smoke and tear gas that we could not see where we were running, while snipers were positioned on top of our homes ready to strike anything that moved. As hard as these memories and events were, they strike a mesmerizing chord and will always remain unmatched and unforgettable.

One such event, certainly not the first or last, I experienced as a child in the 80s remained in my memory to this day, when two Israeli soldiers took my new bike, destroyed it and ran it over with their truck (on my very first ride). I was so excited about trying the new bike but as soon as I got to the end of the street, an ugly brown truck (probably the ugliest thing I have ever seen as far as I can remember) stopped next to me and two soldiers jumped out of it with their arms. They stopped me, took the bike, and brutally proceeded to beat me up. I saw unspeakable hatred in their eyes and realized the extent of the calamity that had befallen us as Palestinians. They took my bike and threw it under the ugly truck. I knew what they indented to do… so I did not back down. I kept trying to get my bike back, but as soon as I turned to the bike the two soldiers resumed to brutally beat me using the back of their weapons to dissuade me from my purist to regain my bike. It was to no avail. The strangest thing in this whole encounter was that I was not afraid of these soldiers or their weapons (for once), because I knew what they were doing was very wrong (in the very simple and childlike sense of the word). The source of my pain then was not the fact that they were beating me up, but the fact that I was unable to regain what was mine. I felt the loss of ownership of my bike which as a child was valuable to me; a small taste of what my ancestors must have felt when they were forcefully driven out of their won homeland… and it was awfully painful to say the least.

Although it was not the first or last encounter of its kind, I will never forget this particular incident and will always recall how troubled these grown men were as I saw the profound fear in their eyes even though I was just a little child struggling to free my bike which I had to watch them crush at the end. That day I returned home with my new bike being carried on the shoulders of those who witnessed the incident and stunned by my persistence to get it back. I also returned home with something far more profound than that—a painful memory; a new me and a new outlook at our struggle. This event gave me a new purpose and aspiration in life that is much bigger than me. A new meaning and new realization was truly ingrained in my heart and mind that we have a cause, and a just one… that violence can only breed violence… that this cyclical narrative of the conflict will live on until peace and justice prevail and what has been taken by force is returned to its rightful owners… that we must realize the necessity and urgency of finding a satisfactory and just settlement to put an end to this occupation, and proactively act on it.

Illustration rendered by Michael N. Erickson, 2018

My experience in the camp also brings memories of coping with water shortages as one of the daily ordeals imposed on us by our occupiers who enjoyed the riches of our land. As farmers, we had a special tie to our farms and a strong bond to our land. Irrigating the land was viewed as an utmost duty to be sought, though insurmountable, considering the intermittent water supply that we had to deal with. The ongoing water shortage that persists until present day cannot be simply attributed to droughts or infrastructure mishaps. It was simply a matter of supply, and a way of exercising control over the Occupied Palestinian Territories by weaponizing water. This particular experience shaped my view of water not only as an effective weapon that can be deployed to assiduously put pressure on others in order to achieve political or strategic gains in the process, but also as a key to potentially resolving disputes. The weaponization of water was truly shocking considering that it is a basic human need, but by the same token an eye-opener of its importance and impact that it could potentially have. This is because water is life. This is an immutable universal truth. Nothing epitomizes the meaning of life itself more than water. Water is needed by everyone and for many uses including subsistence uses. For this reason, water is regarded by many as a generator of conflict and a likely source of war if not managed well. However, in my previous book, *Water as a Catalyst for Peace*, I turned this idea on its head by making the case that water can and should be viewed and treated as a venue for cooperation rather than *casus belli*, reflecting on my own experience growing up in Gaza. Based on scanning the global landscape of water related treaties, I argued that equitable allocation of water not only can help in building trust between disputing nations, but it can also set the stage for other major issues, such as land, sovereignty, and borders to be amicably resolved. The bottom line is that water can bind people together instead of dividing them. Raised during the Palestinian *Intifada* (uprising) and growing up in a world of environmental inequity, I recognized the value of water as the sustainer of life and peace early on in my life. This recognition of the value of water drove my passion that shaped both my life as well as my career aspiration, both of which exemplify a relentless quest for pragmatic ways to bring peace to disputing nations globally, including the Israeli-Palestinian conflict, one that is near and dear to my heart. The underlying paramount premise of my research as reflected in this book is that water itself can and should be used as a catalyst for peace to cement comprehensive and lasting agreements ending years of hostility and bloodshed.

When it comes to the Middle East, often times, water and war are interchangeably used in many contexts. As an extension to the land in dispute, water *ipso facto* constitutes a fundamental ingredient in the making of the Middle East geopolitical discourse and often discussed in conjunction with state sovereignty and international borders (Abukhater 2009). As is the case in many regions around world, conflict over water resources allocation between the Israelis and Palestinians occupies a center stage in their ongoing conflict, and remains one of the most severe obstacles confronting the entire region. Both countries need water for their survival, but one side is using more than just their fair share. In addition, in this semi-arid region water is not always abundantly available to begin with. Because of that, many tried to make the argument that water is the source of the Palestinian-Israeli conflict, frequently referencing the Six-Day war of 1967 in which Israel took control of major water resources in the region. However important the role of water as a driver of conflict and peace is in this region, this conflict is *not* about water. Water is critical but ultimately the two sides fight for control of land, independence, and security, among other things. The conflict has many weighty dimensions including ideological, political, and environmental dimensions that need to be taken into consideration and fully understood as a prerequisite for any future resolution to be successful in reaching an acceptable negotiated settlement by both sides.

As a refugee myself, having witnessed the daily struggle that the occupation imposed on us and the first Palestinian *Intifada* (uprising) as well as the ensuing peace talks, I sometimes struggle to see an end to this conflict, considering the way it is perennially framed and therefore dealt with. Having endured the pain of statelessness and the ongoing pursuit of statehood, peace and stability, I have a unique and important viewpoint that has been shaped by my real personal and professional experience living in refugee camps for most of my life and most recently in diaspora. This experience that forced us as Palestinian refugees to form refugee camps in an attempt to survive, as painful as it might be, gives me the ability (and most importantly credibility) to reflect on geopolitical events and any future resolution proposal with pragmatic measures. This also allows me to put things into real perspective and challenge unrealistic or out-of-touch approaches, many of which are being considered today. There has been a sense of belonging that came with being a Palestinian refugee that anchors all Palestinians living inside the occupied territories or abroad. This sense of one community bonds us together with a shared loss, but most importantly with a common purpose and goal of statehood and self-determination. For us the story of Palestine is quite simple—displacement, appropriation, confiscation of land and resources, and continued occupation. However, the conflict, which is usually portrayed as very complex and nuanced by design, still continues despite the many aforementioned attempts to resolve it. Many groups claim ownership of the land of Palestine and its significant sites on theological or ideological basis. Others seek to secure access to its land and what comes with it in terms of natural and physical resources. The Palestinians, for example, most of whom were dispossessed and displaced from their own land, seek peace, security and an independent state of their own. This, although sounds simple, represents a huge dilemma for the very foundation of the state of Israel, who displaced them to begin with and continues to this day to occupy a large part of their land and whole identity that has been undermined by the Israeli public discourse. This flawed public discourse was powerfully reinforced by their anxiety-driven quest for "security" and their proclaimed existential angst that drives most of Israel's domestic and foreign policies when it comes to their relationship to the Palestinians (Khalidi 2013).

The Palestinian people have consistently been denied the right to self-determination and just and lasting peace that could comprehensively and satisfactorily put an end to their suffering. Any recognition of their rights as refugees automatically means that Israel must face the harsh reality of having to deal with meeting its legal obligations to end occupation, allow Palestinians to return back home and provide compensation for those who do not wish to do so. All of these options pose a real issue for Israel and therefore are not even fathomable as real solutions in the minds of its current leaders. However, to reach peace, compromises need to be made, and not just from the Palestinian side only. Compromises as well as gains must be reciprocal in a win-win resolution. Creating a cadence of accountability calls for an immediate halt of illegal Israeli settlements, which have been repeatedly and widely criticized as being flagrant violation under international law, and finding a resolution for the Palestinian refugees forced out of their home in 1948, 1967 and beyond. Any impactful solution must recognize and deal with these central issues, which cannot be viewed as compromises but rather equally critical legal obligations. Otherwise, the results will not produce a much better reality for both sides of the divide than what the current situation holds for them.

It is important to point out however, that this book is not written for the purpose of documenting or presenting historical events or arguing for one solution or another. Neither is it written about my life nor the lives of many Palestinian refugees in Palestine or in diaspora. This book is written with the intention of exposing past events that led to the current situation, evaluating the current state of the conflict in light of new circumstances and the reality on the ground, and exploring viable options for a practical and satisfactory negotiated settlement that ensures justice and viability.

Denver, USA Ahmed Abukhater

Acknowledgements

It is my great pleasure and privilege that I had the opportunity to write this book, where I am able to offer a transformative framework for practical conflict resolution in the context of the Palestinian-Israeli dispute, reflecting my own firsthand experience living in Palestine. I owe special thanks to Margaret Deignan and the team at Springer for the invitation to bring this book into fruition. Special thanks go to the talented Michael N. Erickson for his artwork and sketches he contributed to this book.

I greatly appreciate the time and feedback of all those who encouraged me to pursue this work, especially my parents, who planted the seeds of inspiration for the making of this book with their endless love and thoughtful feedback without which this book would have been incomplete. I am extremely indebted to their unconditional love and unwavering support.

Contents

Acronyms and Abbreviations

ADR	Alternative Dispute Resolution
ADZ	Armistice Demarcation Zone
BAR	Basins at Risk
BATNA	Best Alternative to A Negotiated Agreement
BDS	Boycott, Divest, and Sanction
BSG	Black September Group
CBM	Confidence Building Measure
CIA	Central Intelligence Agency
DOP	Declaration of Principles
DSM	Demand-side management
EAP	Environmental Action Plan
FAO	Food and Agriculture Organization of the United Nations
GDP	Gross Domestic Product
IBWC	International Boundary and Water Commission
IDRC	International Development Research Center
IFR	In-stream Flow Requirement
ILC	International Law Commission
JFF	Joint Fact Finding
JWC	Joint Water Committee
LHWP	Lesotho Highlands Water Project
MCM	Million Cubic Meters
NGO	Non-Governmental Organization
OPT	Occupied Palestinian Territories
PASSIA	Palestinian Academic Society for the Study of International Affairs
PFLP	Popular Front for the Liberation of Palestine
PISGA	Palestinian Self Governing Authority
PLO	Palestine Liberation Organization
PNA	Palestinian National Authority
PWC	Permanent Water Commission
SNT	Single Text Negotiation

SSM	Supply-side management
UN	United Nations
UNDP	United Nations Development Program
UNEP	United Nations Environment Programme
UNESCO	United Nations Educational, Scientific and Cultural Organization
UNICEF	United Nations Children's Fund
UNRWA	United Nations Relief and Works Agency
US	United States (of America)
USA	United States of America
WHO	World Health Organization
ZOPA	Zone of Possible

List of Figures

Chapter 2

Chapter 3

Chapter 4

Chapter 5

About the Author

Ahmed Abukhater An award-winning author and TED speaker, Dr. Ahmed Abukhater is an architect, environmental scientist, and an urban and regional planner by trade. Raised during the Palestinian Intifada (Uprising) in the Gaza Strip, Dr. Abukhater recognized the value of water as the sustainer of life and peace. His life and work are dedicated to finding pragmatic ways to bring peace to disputing nations globally, including the Israeli-Palestinian conflict. He is the author of "Water as a Catalyst for Peace - Transboundary Water Management and Conflict Resolution" and currently leading Product Management and Innovation at The Boeing Company. Throughout his career, he served in a variety of senior and executive positions in both governmental and nongovernmental organizations, including Senior Director of Mapping and GIS at Trimble, Global Director of Product Management at Pitney Bowes Software, Esri's Global Industry Manager for Planning and Community Development and Director of PLACES (Planning, Architecture, Construction, Engineering and Sustainability), and GIS Director at the University of Texas at Austin. Dr. Abukhater also served as an Assistant Professor of Geography and Planning at the State University of New York at Albany and at the University of Texas at Austin, USA. His experience ranges from the technological and practical aspects of planning and community development to public policy and implementation matters. Dr. Abukhater is a noted expert in a number of areas including geospatial and navigation solutions, strategic marketing and product management, sustainable community development, and trans-boundary water resources management and conflict resolution. His research examines the social, economic and environmental factors that drive the 21st century urban development, geopolitical discourse, resource access and allocation, and conflict prevention and resolution. He holds a Ph.D. in Community and Regional Planning from the University of Texas at Austin, Executive Education in Public Leadership from Harvard Kennedy School, Executive Management and Leadership from the MIT Sloan School of Management, a dual Master's degree in Urban and Regional Planning from the University of Illinois at Urbana-Champaign, and a Bachelor's degree in Architectural Engineering. Throughout his career, Dr. Abukhater has authored numerous publications, served on many governing and advisory boards, and received over 20 prestigious awards for his work.

Abstract

When it comes to the Middle East, often times, water and war are interchangeably used in many contexts. As an extension to the land in dispute, water *ipso facto* constitutes a fundamental ingredient in the making of the Middle East geopolitical discourse and often discussed in conjunction with state sovereignty and international borders. As is the case in many regions around world, conflict over water resources allocation between the Israelis and Palestinians occupies a center stage in their ongoing conflict, and remains one of the most severe obstacles confronting the entire region. However important the role of water as a driver of conflict and peace is in this region, this conflict is *not* about water. Water is critical but ultimately the two sides fight for control of land, independence, and security, among other things. The conflict has many weighty dimensions including ideological, political, and environmental dimensions that need to be taken into consideration and fully understood as a prerequisite for any future resolution to be successful in reaching an acceptable negotiated settlement by both sides.

The pressure that has been building is a byproduct of the peace process that resulted in a situation where the Palestinians feel no meaningful resolution has been reached. Throughout the years following the Oslo agreements, things went from bad to worse. More land appropriation, more settlement, siege and multiple wars resulted in an isolated Gaza and fragmented West Bank. It is time to look back and assess why these agreements did not produce the outcome it promised. It is time for us to look also for ways out of this disparate situation where justice and fairness can be achieved as the basis of a viable and lasting resolution. This is precisely what this book intends to accomplish. This book is written with the intention of exposing past events that led to the current situation, evaluating the current state of the conflict in light of new circumstances and the reality on the ground, and exploring viable options for a practical and satisfactory negotiated settlement that ensures justice and viability. It attempts to draw lessons and conclusions, based on the methodology outlined in my previous book, *Water as a Catalyst for Peace*, and further will chart the course to a more practical framework of achieving regional stability and justice. Water resources allocation remains at the heart of this framework.

Prelude—Level Setting

Water for Peace: Extinguishing the Fire of War and Conflict

Photo TEDx Boulder, 2016

I grew up in the Gaza Strip and like many of you, I took water for granted until one day I turned on the tap and there was no water for us to drink. Shortly after that the first Palestinian uprising/known as the *intifada* erupted. It was a turning point in my life that got me interested in water issues and fueled my passion about finding solutions to the global water crisis. It also got me to think about what would happen if we do not have enough clean water for everyone to drink.

You would not drink this glass of water if I told you there is a 90% chance that it is unfit for human consumption, would you? Of course not. But what if it is all you had? What would you do?

This is the exact ethical and humanitarian dilemma facing my family and more than one million other human beings in the Gaza Strip tonight, teetering on the edge of one of the biggest humanitarian disasters on the planet.

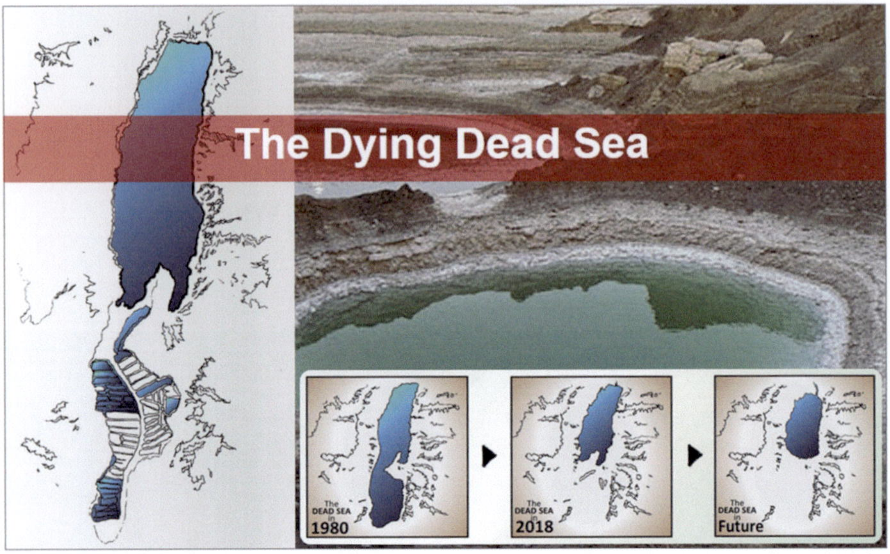

Source The Dying Dead Sea. Illustration rendered by Michael N. Erickson, 2018. *photo credit* FOEME

According to the World Health Organization, 90% of the water in Gaza is unfit for human consumption. Access to adequate water in terms of quantity and quality is severely restricted. The Jordan river is drained and polluted, and because of that, the Dead Sea water level is dropping one meter each year (three feet each year) creating these massive sink holes that you can see on the screen.

It is estimated that within the next 3 decades the Dead Sea will live up to its own name and completely vanish, if nothing is done to save it.

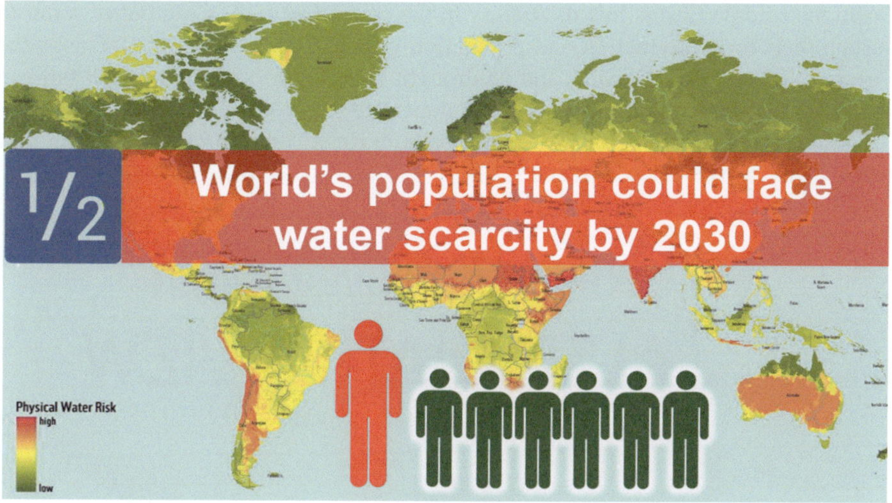

Source TEDx Boulder, 2016. *Photo credit* FOEME

By the way, this region is not an anomaly; there are many regions around the world also facing serious water shortage issues, as you can see on the map. This problem is so pervasive that 1 out 7 people in the world does not have access to clean drinking water. And it is getting even worse. Half of the world's population could face water scarcity by the year 2030.

Source TEDx Boulder, 2016 WWF Water Risk Map, 2016

So the Middle East is just the canary in the coal mine; if they run out of water to sustain their food production and population growth, then we are next! This is the Colorado River (Lake Powell) and its drought sounds all too familiar. Ask farmers in California and people whose lives depended on what once used to be the Colorado River Delta. You can see the "bathtub rings" on the walls indicating past water levels as they slowly receded.

Source TEDx Boulder, 2016. *Photo credit* Michael Melford, National Geographic Creative

Today we are being challenged on a whole new biblical scale and it is a stark reminder of what's at stake here—our survival. So let's talk about how we ended up in this situation, and the implication of not having enough clean water for everyone to drink, which to some means water wars. The Former UN Secretary General Mr. Kofi Annan once said "Fierce competition for fresh water may well become a source of conflict and wars in the future." And Mr. Ismail Serageldin, Former Vice President of the World Bank said: "The wars of the twenty-first century will be fought over water."

Nothing epitomizes this view more than the term rivals. It actually comes from the Latin word *revalis*, which means one who uses a stream in common with another. This water war narrative is widespread and focuses on the demise of humanity as a result of water shortage and potentially water wars. And what this narrative is pretty much saying is that water was the source of conflict in the past and will continue to be the source of conflict and even outright wars in the future. And the Middle East is going to be the first stage on which these water wars will play out. Although very interesting and pervasive, this water war narrative is non-factual, scientifically inaccurate, and fails to offer real solutions or future direction beyond inciting fear.

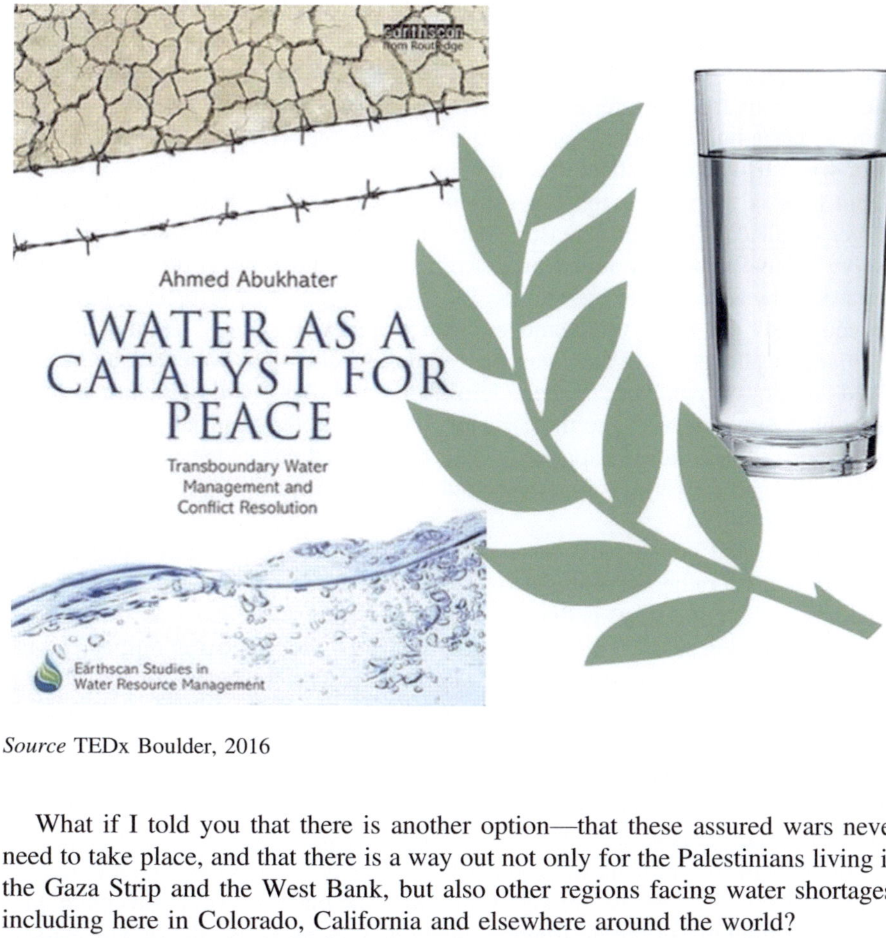

Source TEDx Boulder, 2016

What if I told you that there is another option—that these assured wars never need to take place, and that there is a way out not only for the Palestinians living in the Gaza Strip and the West Bank, but also other regions facing water shortages, including here in Colorado, California and elsewhere around the world?

We looked at many conflicts in the past as well as negotiated settlements and we found no evidence to substantiate the water war argument. On the contrary, we discovered that when water was involved as part of the dispute, people tended to cooperate not altercate. And the closest we ever came to a water war is this (a picture of kids playing with water)—it's not so bad, it's actually a lot of fun!

There are many examples from all over the world, India and Pakistan, Lesotho and South Africa, where these countries entered into negotiation over water while they were in an officially declared state of war. However, as they set out to resolve the dispute over water they found themselves resolving a lot more than just water.

Water gave them the courage to see eye to eye and negotiate and resolve other high politics issues; seemingly insurmountable. Water also allowed them to consider other important dimensions linked to water like land, sovereignty, food security, and even energy (that's because you can trade water for other things like food production or energy production—and that's what we call virtual water).

But then how do you explain this? This is what over 20 years of negotiation did—war and destruction. This is Gaza City after three brutal wars in only 6 years. And it is yet another stark reminder of what happens when we choose to ignore one the most basic human rights—water. And it is easy for people to confuse this with water wars. It would be shocking if they do not.

But here is the reality; the Palestinian-Israeli conflict is widely known as the mother of all conflicts for a reason—and that's because all previous attempts to resolve it have miserably failed or exacerbated the situation on the ground. You may ask: how could peace negotiation for over 20 years lead to more wars? Well, are you ready for this? We found out that water was actually left out of these negotiations altogether. Water was one of three issues that were not addressed, talked about, or resolved. Water, Jerusalem, and the right of return for the Palestinian refugees were the three issues that were deferred to the final round of negotiations, which never happened. They overlooked water, they overlooked the right to life, and the catalyst for peace, and this is the result—wars, destruction and regional instability.

We also found out that like many other disputes, water was not the source of their conflict, but rather an obstacle to peace, and that was the bad news. And by the same token, the good news is that if we remove this obstacle then we are one step closer to achieving peace. This is a remarkable seed of hope and a stentorian call for peace. What we need to do is to use water as a catalyst for peace by allocating it equitably, and then and only then will we be able to build solid foundation for a just, lasting and robust peace.

If you can negotiate over water, the sustainer of life, you can negotiate over anything, because what you are saying is that you value life… all lives… not just yours, but your rivals as well. When you offer someone water, you offer them life; you offer them the ultimate olive branch. Obviously not the questionable cup of water I offered you at the beginning of this talk. And once you get to this point, you will not be thinking in the destructive binary way of us versus them. You will be thinking about one thing—how can we collectively work together to elevate and advance our lives and the lives of others.

For me water is the ultimate peacemaker because it catalyzes peace not war. And if we truly understand what this really means and take action in our own lives, and take ownership of this movement that focuses on conserving resources not squandering them, sharing them not dividing them, building bridges not walls, and creating allies not enemies, we will be able to change not only the world in which we live, and which we will pass onto our future generations, but also the lenses through which we see the world and perceive it—either a place to fight or a place to live in peace. Let us fight for peace not water and use water to resolve our fights. Because water after all is used to extinguish fires not ignite them.

Transcribed from TEDxBoulder Talk. This talk was given at a TEDx event using the TED conference format but independently organized by a local community. Learn more at http://ted.com/tedx.

Chapter 1
Historical Context—Tracing the Origins of the Problem

A lie told once remains a lie but a lie told a thousand times becomes the truth.

—Joseph Goebbels

Water is important but it is not why Israelis and Palestinians are fighting. Water is an obstacle to peace and removing this obstacle can bring them closer to achieving peace. There are other key issues driving and sustaining their dispute and this chapter will trace some of these key root causes of the conflict between Israel and Palestine. It will also provide a brief historical overview of the origins of the problem starting with the Zionist movement and the Jewish immigration to Palestine at the onset of the 20th century, all the way to the establishment of the State of Israel in 1948 and the ensuing events leading to the Six-Day War of 1967 in which Israel expanded its territories. This chapter is not intended to provide comprehensive historical overview of the conflict or the countries involved in it. There are many books that are concerned with the history of the region, and this book is not one of them. This chapter will cover key historical events to give the reader relevant and impotent contextual background necessary for the understanding of the current events and future direction. It will also expound on the current situation and the modern events such as the Gaza siege and recent wars fought in Gaza, the most recent Palestinian *Intifada* (uprising), best known as the *Third Intifada*, and how these events impacted each side and shaped the perception of the prospect for peace and its attainability in this region. There will be a particular focus on how these local, regional, and global geopolitical events shaped, and in some cases changed, the map of historic Palestine, as illustrated in Fig. 1.

Palestine; a land without a people for a people without a land. This is the lie that was perpetrated about the land of Palestine for many decades to encourage and justify increased Jewish immigration and later the invasion of the country and the displacement and disposition of the Palestinian people from their homeland in 1948 and later in 1967 and the ensuring tragic events throughout the past century. During these momentous events, the original inhabitants of the land were violently and forcefully removed from their ancestral homeland to make way for the new comers who often justify their position based merely on mythological grounds. The dispute over the historic claim of the land between the Mediterranean Sea and the

A. Abukhater, *Palestine - Peace by Piece*,
https://doi.org/10.1007/978-3-030-03985-1_1

Fig. 1 Palestine; yesterday and today

Jordan River is one of the bloodiest and most devastating conflicts in the world. The significance of this land to all three major religions ipso facto made it a unique place that all three religions often vie to secure access to Jerusalem and its surrounding areas. This part of the world has witnessed a number of momentous events in which different groups battled to gain control over the land and its resources. Still casting their dark shadow on the present political scene, these events that are engraved in the hearts and minds of the Palestinian refugees shaped the region's history, character, and therefore claims of ownership. The events of 1948 that led to the establishment of Israel, which triggered armed conflicts that lasted for decades, are referred to by the Palestinians as the *Nakba* in Arabic or the "catastrophe" (Shamir and Haddadin 2003).

The Palestinian tragedy however did not actually start with the war of 1948, although it was one of the most far reaching geopolitical events. It has its roots over 200 years ago with significant political and diplomatic events that paved the way for what conspired in 1948 and beyond. It all started with Napoleon Bonaparte's invasion of Palestine in 1799 and his call on the Jewish people around the world to revive the idea of a Jewish state in Palestine. Napoleon Bonaparte was defeated and his campaign ended but not his idea. His aspiration of establishing a Jewish state under colonial sponsorship in Palestine remained alive and well. Forty years later, this idea was resurrected by Britain whose main goal was to safeguard its strategic interests in the region as a colonial power. Despite the very small number of Jewish people in the Ottoman-controlled Palestine at that time that did not exceed three thousands, Britain intended to open Palestine to become an immigration destination and ultimately a Jewish state. In the early 1900, the idea of statehood had firmly took root around the world, which gave rise to the idea of a Jewish state in Palestine. In 1897 the Zionist movement, which was established as a political organization under Theodor Herzl, envisioned and lobbied aggressively for the establishment of an independent Jewish homeland in Palestine, including Trans-Jordan. The Zionist plan was to ultimately establish an exclusively Jewish homeland in Palestine in terms of its identity socio-political structure and ethnic composition (Pappe 2006). The location of Palestine was under investigation during the early years of the Zionist movement to gage its suitability for a potential Jewish homeland. Their investigation concluded that "the bride is beautiful but she is married to another man," implying that Palestine at that time was inhabited by the Palestinians who had their own communities, a historical fact that effectively refutes the claim that Palestine was a land without a people for a people without a land. Nonetheless, the Jewish people answered the call this time and started to invest heavily in building Jewish settlements inside Palestine, although the country was under Ottoman control at that time.

Britain had a special and instrumental role in the Palestinian-Israeli conflict and the formation of the Jewish state in Palestine. Acting as the colonial sponsor and protector, Britain supported and advocated for this goal to safeguard its interests and pursuits in the region, namely the vicinity to the Suez Canal with its strategic importance. As such, Britain viewed the Zionist movement as a partner and ally in a strategic colonization project of the region that will protect and safeguard its interests and further reinforce its imperial influence in the Middle East. Considering the convergence of its interests with those of the Zionist movement, Britain exhibited an ostensibly favorable position towards the Jewish immigration and the depopulation of Palestine. The Zionist movement began to implement its plans in the early nineteen hundred by purchasing significant land north of Palestine and driving its local Palestinian inhabitants out to make room for the new Jewish immigrants (as shown in Fig. 2). As a result, large numbers of Jews from all over the world started to move to and settle in Palestine. This was clearly a different form of colonization because its intention was not just to take advantage of the resources of the region by controlling the local inhabitants but rather completely depopulating the region by forcefully displacing its original owners. It is critical to point out that this goal has not changed much since then and its resonance is clearly evident today in Israel's current policies that by design denies the Palestinians the right to return to their land from which they were forced out.

These early events that brought a new demographic group to the area had dire consequences for the local Palestinian communities that lived in Palestine at the time. These Arab communities were cognizant, albeit not fully aware of its serious consequences, of the Zionist movement and its plans to leverage religious claims to occupy Palestine and ultimately establish a homeland with a new and exclusive Jewish identity. This identity was viewed by the Palestinians as a poisonous dagger in the heart of the Arab region. Although determined to establish and reinforce its roots in Palestine, the early leaders of the Zionist movement realized that the time was not conducive to establish their Jewish state in the early 1900. Instead, they supported the British mandate to pave the way for the establishment of the foundation for their future state in Palestine. In a secret agreement signed on May 16, 1916, notoriously known among Palestinians to this day as the Sykes–Picot Agreement, Britain and France mutually agreed to divide the region with Palestine under Britain control, in anticipation of the defeat of the Ottomans in World War I. World War I ended in

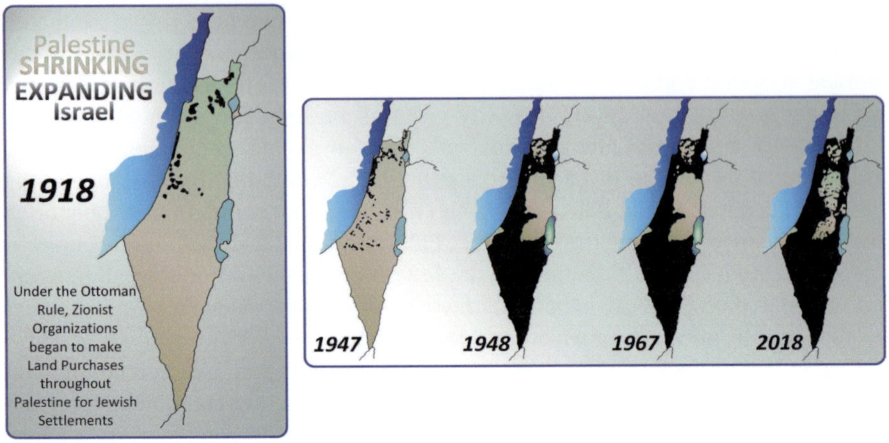

Fig. 2 A map of Palestine in 1918. *Source* Illustration rendered by Michael N. Erickson, 2018 (adapted from Visualizing Palestine)

1918 and turned their prophecy into a reality, allowing for the actual implementation of the Sykes–Picot Agreement.

As a result of this agreement, Palestine came under the control of Britain in 1920 which lasted until 1948, opening the way for the creation of the Jewish state. This was a pivotal moment in the history of the region that spawned a century of resentment, conflict and misery. The disastrous results inherited by the Middle East and its inhabitants in the form on redrawing the region's political boundaries with little regard for national, ethnic or religious identities, provoked widespread discontent. The agreement was problematic not only for the Palestinians but also for the entire Middle East. The agreement came as a sordid betrayal to the British promises of freedom to the Arabs in exchange for their support against the collapsing Ottomans. It was also in direct contradiction with the U.S. President Woodrow Wilson's vision of self-determination for the nations that were under the control of the Ottoman Empire. The legacy of the Sykes-Picot Agreement has long been criticized and explicitly assailed by many Arab nations including the Palestinians who clearly saw it as an invasion of their land and by extension sovereignty. For the Palestinians, the Sykes–Picot Agreement personifies the spark of a long and brutal confiscation and grinning occupation of their land because of its influence in establishing the foundation of the Jewish state, and the seeds for the long term instability of the Middle East as a whole.

In 1917, Arthur Balfour, Britain Foreign Secretary, wrote a letter on behalf of his country promising the Jewish people a national homeland in Palestine. This letter known as the Balfour Declaration (or the *Promise*) is loathed by the Palestinians until today, due to its long-term consequences that made the establishment of the Jewish state a reality and further triggered the chronic state of conflict between Arabs and Jews throughout the entire region and most recently morphed into the current

Palestinian-Israeli conflict. Ilan Pappe, a prominent Israeli historian and professor, in his book, *The Ethnic Cleansing of Palestine*, writes:

> The moment British foreign secretary Lord Balfour gave the Zionist movement his promise in 1917 to establish a national home for the Jews in Palestine, he opened the door to the endless conflict that would soon engulf the country and its people.

As such, the Palestinians referred to it as the promise made by those who do not own to those who do not deserve, alluding to the fact that Britain had no legal or ethical right to promise Palestine to anyone, much less to the Jewish people whose population did not exceed 5% of the total population of Palestine at that time. The end of the Ottoman Empire and the start of the more sympathetic and supportive British Mandate over Palestine ushered in a new era for the Zionist movement which saw and capitalized on the opportunity to accelerate the transformation of Palestine into a Jewish state. During the British Mandate of Palestine, there was a mutual understanding between the Zionist movement and Britain that the latter would act to prepare Palestine politically, administratively and economically to allow for the establishment of a Jewish homeland (*or Erez Israel*). The crucial role that the British played in enabling and accelerating the Jewish immigration and the appropriation of Palestinian land to accommodate the expanding Jewish presence cannot be overstated. The *modus operandi* of the British Mandatory authorities in Palestine was primarily in favor of the establishment of Israel on Palestinian lands, by safeguarding and enabling the Zionist movement to acquire as much land as possible pre-1948. Protecting and enabling the Jewish Agency, the British Mandatory authorities implemented draconian measures and systematic discriminatory policies against the Palestinians who were distortedly viewed and treated as an indigenous population in need of control, while considering the Jewish people as a legitimate representation of a larger group in need of independence. As such, Britain aided the Zionist movement in land acquisition and the establishment of their own infrastructure, which in turn transformed the Zionist project from an idea into a concrete reality.

Frustrations were compounded by this evident British partiality towards the Jewish people in Palestine, demonstrated by the Balfour Declaration, coupled with the proactive attempts of the Zionist movement to lay claim to the Al-Buraq Wall (Wailing Wall). This flashpoint caused widespread Palestinian resentment that led to riots and conflicts, known as the Al-Buraq Uprising. Tragic but predictable clashes between the two groups and with the British authorities erupted. Despite the British brutality in suppressing these demonstrations, Palestinian riots and clashes with the Jews escalated throughout the country in the early 1930s and led to the killing and imprisonment of thousands of Palestinians.[1]

A new national Arab identity started to take shape across the Arab world and the Palestinians were at its heart. In 1936, cognizant of the implications of intensifying Jewish immigration to Palestine, Palestinians went on a general strike that lasted for six months until the British government changes its one-sided policy that was

[1]It is estimated that five thousand Palestinians were killed by British authorities between 1935 and 1939 and more than 15,000 were wounded (Aljazeera 2003).

widely viewed as an enabler of this Jewish immigration. This was the first sustained revolution by Palestinians in which Arabs from other neighboring nations were also mobilized. Britain once again met this rebellion that shook its foundation with brutal crackdown that included the killing and imprisonment of thousands of Palestinians and demolition of Palestinian homes as a collective punishment to deter others from participating in the strike.

Following the six-month-long Arab general strike, the Peel Commission was appointed in 1936 to investigate the unrest and provide recommendations to restore calm. The Royal Commission (Peel), recommended a partition plan in 1937 that would create two areas; the northern coastal area for the Jewish people and another for the Palestinians in the southeast side along the Jordan River to be potentially combined with Trans-Jordan, as shown in Fig. 3.

This plan, which suggested the "transfer" of Palestinian population from the areas allocated for the Jews, was uniformly rejected by both the Palestinian community and Arab leadership as the idea of a Jewish state on Arab land was strongly opposed. The idea of "transfer" of Palestinians out of their land to make room for the new Jewish immigrants was a critical success factor for the Zionist movement but equally devastating for the Palestinians. This idea implied a potential systematic depopulation of the Palestinian communities to reinforce the colonial most prominent lie about Palestine being a land without a people for a people without a land.

Armed Palestinian resistance continued, so did the Jewish gang attacks against Arab targets, coupled with the continued British heavy-handed policies in crushing the Palestinian revolution and their leaders. Above all, Palestine remained an undivided country. British authorities were anxious to quickly eradicate any immerging Arab revolts or any voices in opposition of their policies or the presence of foreign influence in Palestine. Britain ultimately succeeded in crushing the revolution and getting rid of the Palestinian leadership, most of whom were either executed, imprisoned or forced to leave the country. Given the leadership vacuum this created for the Palestinians, historians believe that this power vacuum had detrimental impact on the Arab communities in Palestine, contributing to their *Nakba* and eventually paving the way for the establishment of the Jewish state of Israel in 1948 (Pappe 2006).

A sequence of global events, namely World War II, decisively proven the significant role of oil in expanding global influence and control. As such, the United States sought the alignment with Saudi Arabia (the world's largest oil reserve) which in turn was also cognizant of the rising role of the U.S. as a superpower in the global scene. As the U.S. became more influential in the Middle East geopolitical arena, President Roosevelt promised the country's monarch 'Abd al-'Aziz ibn Sa'ud that the U.S. would not act in Palestine in a way that would be "hostile" to the Arabs or without first consulting with the Arabs and the Jews (Khalidi 2013). This promise however was later abandoned by the U.S. whose interests were primarily shaped and driven by the Zionist agenda to establish a homeland in Palestine.

With the start of World War II and the persecution of the Jews in Europe, hundreds of thousands of Jews fled to Palestine seeking refuge from the brutality of the war taking place in Europe. This mass immigration started to change the character of the demographic making of Palestine and planted the seed for the Palestinian-Israeli

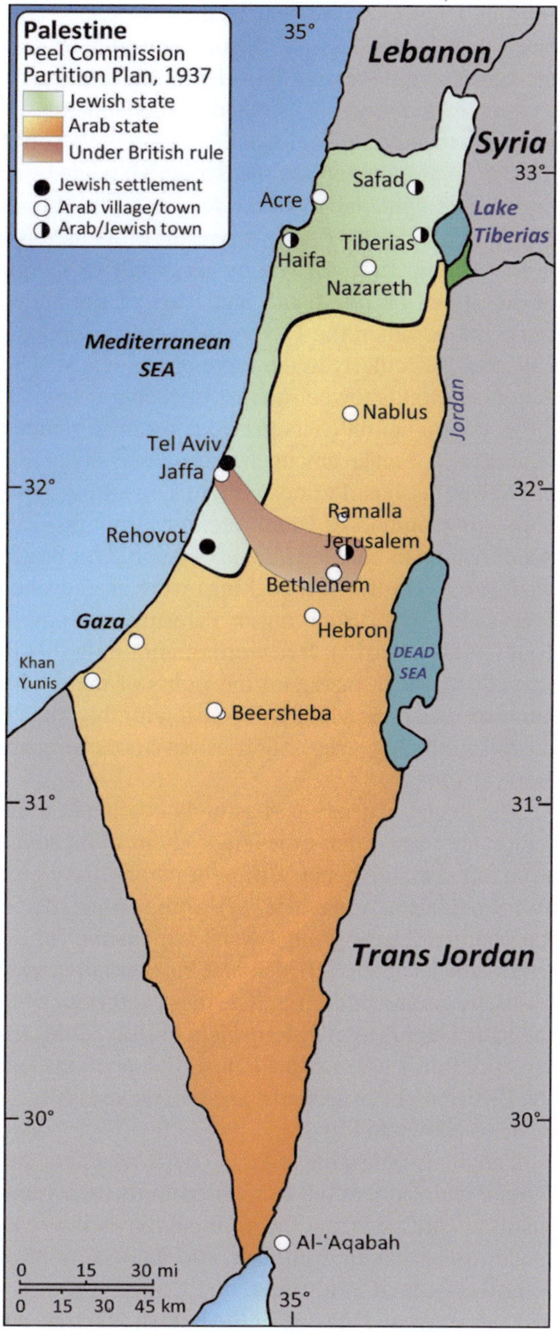

Fig. 3 Partition plan proposed by the Peel Commission report, 1937. *Source* Illustration rendered by Michael N. Erickson, 2018 based on Data from Encyclopedia Britannica, Inc., 2010

conflict. Prior to World War II, the U.S. had very little impact on the events taking place in Palestine. As the Second World War drew to close, the U.S. emerged as the leader of the Western World, quickly aligning itself with the ethos of the Zionist movement and showing a great deal of sympathy. The Zionist movement too realized and capitalized on this global change in power structure and sought alignment with the U.S. in order to replace the British Mandate by a national Jewish state in Palestine as the geographic nexus of the Jewish faith. The role of the U.S. in the region grew exponentially in the 1940s as a result of its increased influence and strong presence in the Middle East. Driven by its desire to safeguard its strategic interests in the Middle East, on one hand, and fears of not showing enough zeal for the Zionist cause on the other, the U.S. was fervently supportive of the Jewish immigration to Palestine particularly in the wake of Holocaust. The U.S., under President Truman's administration, showed interest in seeing a Jewish state in Palestine, which was in direct contradiction to its previous promise (made by President Roosevelt to Saudi Arabia) to not make any decisions or take any actions regarding Palestine that would be viewed as hostile and without consulting with the Arabs. The United States' unflagging support of Israel, pressure from pro-Zionist groups, and the passivity of Saudi Arabia to hold the U.S. accountable for its aforementioned promise (driven by its desire to secure U.S. backing) were all the sine qua non of the Jewish Agency's success to drive Britain out of Palestine and pave the way for the establishment of Israel (Khalidi 2013). It is worth mentioning that this U.S. policy towards Palestine that allowed it to disregard the rights of the Palestinians, tightly align itself to the interests of Israel, while pari passu with this, manage to safeguard its interests with the oil-producing Arab countries have not changed much since the time of President Harry Truman.

The Jewish Agency was astutely cognizant of these geopolitical dynamics in the region and was quick to leverage them to its advantage. With funding and support from the Jewish Agency, different paramilitary groups (such as the Haganah, *Irgun*, and *Stern,* also known as *Lehi*) were formed, determined to push the British out of the country. These groups were responsible for a series of explosions and violent terror attacks against British and Palestinian targets throughout Palestine, including buses, trains and buildings. The most notorious of these attacks were the bombing of the King David Hotel in Jerusalem on July 22nd, 1946, and the Deir Yassin massacre on April 9th, 1948, against Palestinian civilians leading to many deaths among both the British and Palestinians. Despite these events, the map of Palestine remained the same as shown in Fig. 4.

Caught in *flagrante delicto*, with its one-sided colonial policies in Palestine, Britain realized that the benefits from its involvement in Palestine do not warrant or justify its presence and the political mayhem it created due to its role in the region. Under this escalating military attacks and political pressure, the British sought to refer the issue of Palestine to the United Nations, which in turn adopted resolution 181 that proposed the end of the British Mandate and the Partition Plan of November 29th, 1947. As the matter was turned into the United Nations to handle, both the British authorities and the Zionist movement appear to be in *pari delicto*, yet have

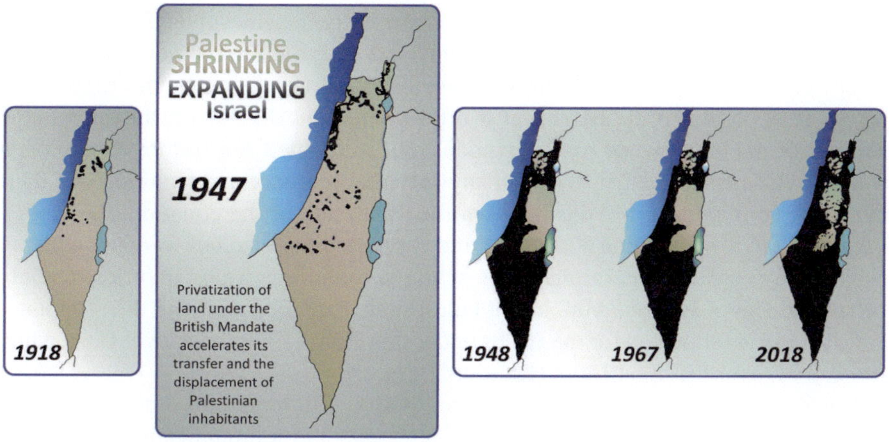

Fig. 4 A map of Palestine in 1947. *Source* Illustration rendered by Michael N. Erickson, 2018 (adapted from Visualizing Palestine)

not been held accountable for the systematic depopulation of Palestine that occurred prior to the sought UN involvement.

Since the onset of the twentieth century, Palestinians as a nation within a pan-Arab identity have been seeking independence and self-determination. Rather than affirming their self-determination by utilizing a more democratic approach to resolving the issue, the newly-formed, unexperienced United Nations instead used another approach by which it proposed a two-state solution on predefined areas, notwithstanding the ethnic and demographic composition of the country. This Plan, which was fully supported and endorsed by the U.S., divided the land into two states; one for the Palestinian Arabs and one for the Jews, with a special international status (*corpus separatum*) for the city of Jerusalem under the administrative authority of the United Nations. Under this Plan, the proposed Jewish state is allocated about 55% of the total land area of Palestine despite the fact that a relatively small Jewish population owned only around 6% of the land, while the Arab state was allocated only 42% of the total land despite the fact that they constituted the overwhelming majority of the population and further owned the vast majority of the land (Söderblom 2003). The Arab state also included the city of Jaffa which was geographically separated from the rest of the partitioned Arab areas.

While this Plan marked the beginning of the end of the British Mandate in Palestine, the Plan was not welcomed by the Palestinians due to its baseless and inequitable allocation of the land and its resources, coupled with the fact that for the first time it gave legitimacy to the Jewish presence in Palestine. With this remarkable Plan, the establishment of a viable Palestinian state was all but doable and practically impossible to implement on the ground due to the clear geographic separation of Jaffa from the rest of the Arab land and Gaza from its farmland. The proposed Arab state was crippled by its noncontagious spatial configuration, in a way echoing the reason why a practically viable Palestinian state is currently unachievable. The Arabs and Palestinians found the idea of sharing their land with the Zionist newcomers from Europe difficult to fathom. Not surprisingly, the Plan was rejected by the Arabs, who viewed it as unfair on the basis that it offered the minority Jewish inhabitants the majority of the land owned by Palestinians, but welcomed and further championed by the Jews who were eager to obtain any sort of acceptance and legitimacy of their increasing immigration to Palestine. On the other side of the equation, the United States and the Soviet Union sponsored and supported the Plan but were not actively involved in protecting its implantation while neighboring Arab countries, namely Jordan and Egypt syphoned the infant Palestinian state by occupying the remainder of the territories allocated for the Arabs under this Plan—Egypt took over Gaza while Jordan took over the West Bank (Khalidi 2013).

The Zionist movement leaders mastered both the military and diplomatic game and deployed both whenever appropriate. Secret talks and communications took place between Britain and the Zionist movement leaders to ensure the viability and success of the Jewish state in any partition plan by securing strategic access to regional water resources. This understating of water as an essential part for the success of their state drove the Jewish territorial acquisition plan to ensure uninterrupted access to vital water resources such as the Galilee area with its abundant water supply.

The failure of the UN and its Partition Plan to adequately provide an acceptable solution to both sides of the conflict was the trigger that set in motion the chaos that ensued which produced the first Arab-Israeli war resulting in the establishment of Israel and the creation of the Rhodes Armistice Demarcation Line marking Israel's borders, including historical Palestine, with the exception of the West Bank and Gaza Strip which were subsumed by Jordan and Egypt respectively (Abukhater 2013). As a result of this Arab complacency and international lack of interest in the Palestinian right to self-determination, only the Israelis obtained self-determination militarily expanding its territory from 55% (as stipulated by the Plan to 78%) while the Palestinians were not allowed to do so (Khalidi 2013). Figure 5 shows a comparison between the 1947 United Nations Partition Plan and the Rhodes Armistice Line. The maps depict that the actual expanded area of the newly formed State of Israel in 1949 significantly exceeds the area that was allocated to Israel under the UN Plan in 1947.

Practically speaking, the Jewish Agency was politically, administratively, and even militarily in charge of Palestine even before the declaration of the State of Israel, due in large part to the British supportive and enabling policies. On the other hand, the Palestinians had no army and Britain outlawed armament of any Palestinian, displaying a very different policy against the Palestinians, who perceived these policies as discriminatory. The results were not surprising with a much larger and well-equipped Jewish Armey made of the *Haganah* and other Jewish gangs compared to a much smaller and ill-equipped group of Palestinian volunteers and the Arab Salvation Army. After all, Britain was seeking an exit out of its messy involvement in Palestine and this Plan gave it what it was looking for—a way out.

On May 15th, 1948, Britain relinquished its Mandate over Palestine and on the very same day the Jewish Agency proclaimed the establishment of the state of Israel, a theocratic state exclusively for the Jews. Through a large-scale and well-executed

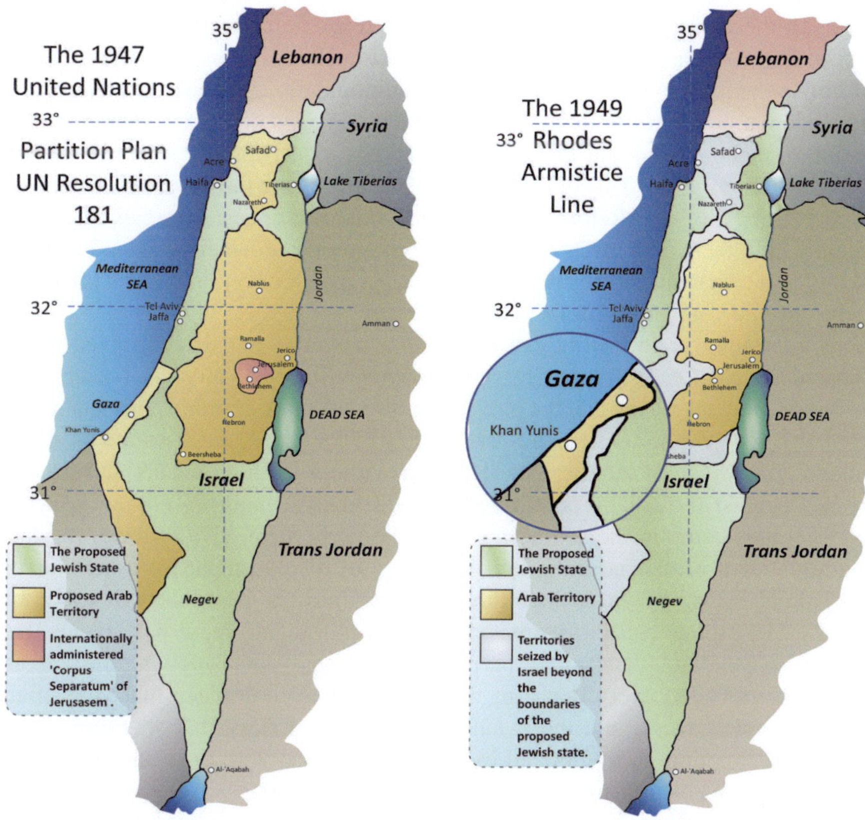

Fig. 5 A comparison between the UN Partition Plan of 1947 and land seized by Zionist militants in 1949. *Source* Illustration rendered by Michael N. Erickson, 2018 based on data from Palestinian Academic Society for the Study of International Affairs (PASSIA)

ethnic cleansing campaign (as described by the Israeli historian Ilan Pappé akin to many others), the Zionist paramilitary gangs waged war against the Palestinians to forcefully drive them out of their homeland to pave the way for the establishment of their Jewish state. Outgunned and outnumbered, the Palestinians and Arab armies involved were no match for the Jewish paramilitary groups that were aided by the British to depopulate the region from its inhabitants. Their goal was to acquire as much Palestinian land with as few Palestinians as possible. This simply meant the dispossession and the displacement of Palestinians. This campaign that utilized the "conquest and destruction" of rural and urban areas of Palestine started as a retaliatory measure in December 1947 but developed into a full-blown campaign in March 1948 that took only six months to execute and resulted in the displacement of more than half of Palestine's native population and the complete destruction of their villages (Pappe 2006). Along with the renaming of these villages, another insidious cover up campaign was implemented to conceal and distort these events and erase the memory of these places from history. In this ongoing campaign, the Israeli narrative has been focused on the denial of the *Nakba* altogether, representing these events as simply a "voluntary flight" of the Palestinian population, despite the clear *prima facie* evidence provided by many historians that suggests otherwise. The intent of this denial of the *Nakba* and the associated events is to belie the forceful transfer and displacement of the Palestinians, which in turn means the automatic denial of the issue of the Right of Return for these Palestinian refugees. This can be clearly summarized in the words of David Ben-Gurion, founder of Israel, who said in 1949 "We must do everything to ensure they [the Palestinians] never do return… The old will die and the young will forget" (Nabulsi 2006).

The hauntingly vivid memory of the *Nakba*, however, remains engraved in the hearts and minds of Palestinians. Every Palestinian village and city has its own story to be told by both the young and the old, generations after generations. This avalanche of evidence both in the written and oral history of Palestine, albeit suppressed and excluded from western media, will remain a witness to these events and a clear *J'Accuse* against those who committed them. I myself, being a refugee growing up in a refugee camp in the southern part of the Gaza Strip, vividly recall my late grandmother gracefully narrating her own story of displacement from our homes in Al-Faluja, taking us on an incredible journey into the past that will never be forgotten. Her stories chillingly opened a clear window into these past events that befell us that year and were further perpetrated in many years following the *Nakba*. She gingerly recalled every detail of those events that took place that year; their location, people's faces, and the meaning of it all in a story that was written by death and destruction, characterized by betrayal, colored with unspeakable horror, and above all was clearly indicative of a horrendous large-scale depopulation and dispossession operation. Never once did she doubt her return to our homeland. The gravity of these crimes committed against the Palestinians proved a priori that cannot be overstated or overlooked. This momentous and painful event gave birth to a new realization among the Palestinian refugees—an aspiration and deep-rooted belief that one day they will return back home. As a result of the 1948 War, the map of Palestine was significantly redrawn as shown in Fig. 6.

The newly established Jewish state was a fulfillment of the Balfour Declaration of 1917, as a result of the 1948 War, taking over not only the land that was allotted to the Jews by the 1947 UN Partition Plan but also 78% of the land allocated to the Palestinians, leaving them with only 22%. The events that took place in 1948 and the actions taken by Israel, in *pari delicto*, have not prevented the UN from asserting and recognizing their claim to an independent Jewish state while denying the Palestinians the right to do the same, as per the Partition Plan of 1947. By so doing, the UN sided with Israel, firstly with its unfair partition plan, and secondly, *pari passu* with this, with its recognition of the newly established Jewish state, which should have been abrogated as the result of its intentionally culpable conduct in the years leading to its establishment in 1948.

It is crucial to note that there was a deliberate campaign of eliminating the remainders of the Palestinian towns and villages, albeit recognized only a posteriori, in an attempt to whitewash and erase any evidence of the Palestinian existence or future

Fig. 6 A map of Palestine in 1948 after the *Nakba*. *Source* Illustration rendered by Michael N. Erickson, 2018 (adapted from Visualizing Palestine)

hope for them to ever return (Hammond 2017). This event significantly, pari passu with the large scale displacement campaign, altered the ethnic composition of the country, creating a major humanitarian crisis. It is estimated that 531 Palestinian villages were completely destroyed and nearly 750,000 Palestinians were uprooted from their homes, becoming refugees in their own land (a list of these villages is included in Appendix A) (Pappe 2006). The ethnic composition of the Palestinians changed from a majority of ninety percent of the population when the Mandate started 30 years earlier down to only two-third of the country at the end of the Mandate. It is no surprise that these events of 1948 were referred to by the Palestinians as the *Nakba* or "catastrophes," due in part to their severe impact on their very existence and identity as a people.

As a result of the *Nakba* of 1948, many of these Palestinians became refugees languishing in refugee camps within Palestine and the surrounding Arab countries. The Gaza Strip came under Egyptian control while the West Bank was under Jordanian dominion. Towards the end of the 1948 Arab-Israeli war, on December 11, 1948, the United Nations General Assembly Resolution 194 was adopted, in which it delineated a resolution for the issue of the Palestinian refugees and asserted their Right of Return (UNDPI 2003). Article 11 stipulates that "refugees wishing to return to their homes and live at peace with their neighbors should be permitted to do so at the earliest practicable date" and "compensation should be paid for the property of those choosing not to return" (UNDPI 2003).

Although this Right of Return for the Palestinian refugees is affirmed by international law, Israel does not allow it, and has actively worked to exclude it from any previous negotiations, as will be discussed in much detail in the fourth chapter of this book. Equality galling, the U.S. never made any serious effort toward a meaningful implementation of this momentous resolution, and ipso facto acquiesced to the Israeli obduracy regarding the issue (Khalidi 2013). The issue of the Right of Return is one of the most significant factors that continue to steer the current conflict between Palestinians and Israelis to this day, and is an issue that needs to be addressed and resolved in any future negotiations between the two sides.

These profound events of the *Nakba* and the forced removal of over 750,000 Palestinians from their ancestral homeland led many of them to either languish under grinding military occupation without basic human rights or flee to surrounding Arab countries living under harsh authoritarian regimes (UNDPI 2003). Under occupation, Palestinians live in miserable conditions in villages and refugee camps without reliable supply of water or security. There has been always a presumption on the part of both Israel and its ally the U.S. about the Palestinians having no concerns about their security, whereas Israel was given a priority and was always portrayed as having serious and deep-seated concerns about their "existential fears." Early on in the conflict, the U.S. and Israel established their discursive hegemony to manipulate narrative, distort facts and corrupt thoughts about the reality and very nature of the conflict, which altogether provided Israel a variety of advantages to act with impunity (Khalidi 2013). The scarily perceptible presence and traces of these Palestinian villages in Israel is a witness to the well-orchestrated operation that took place to eradicate any evidence of Palestinians in order to reinforce the willfully deceptive rubric of Palestine as "a land without a people for a people without a land." Simultaneously with that redemption, was the emphasis on the absence of inhabitance and the depiction of Palestine as hopeless arid land awaiting redemption. The cheer banality of such claims is too obvious to ignore. Palestinians existed inside and outside Palestine for centuries with well-established and distinctive cultural practices and unique national identity.

The habit of considering everything Palestinian as antithetical to Israel represents a huge disservice to the prospect of an overall peaceful settlement and only reinforces the status quo of occupation and marginalization of the Palestinians (Said 2003). This corrupted narrative, along with the intertwined alignment between Israel's position and the United States' foreign policy pertaining to Palestine, created a manufactured reality that made it easy for Israel to depopulate most of the Palestinian villages it occupied in 1948 and continued to indemnify Israel's actions against the Palestinians to this very day. This aligned policy of concealing the reality on the ground that Israel conceived, created and maintained with Palestinians living under conditions that can only be described as inhumane, made it possible for the United States to keep the American public opinion in favor of Israel as a result of keeping it in the dark when it comes to credible knowledge about the conflict. It is precisely this role that the United States played that ubiquitously undermined the Palestinian cause and desiderata in the Israeli-US public discourse and furthermore forced the Palestinian leadership to acquiesce to the political will of Israel and its ally in the ostensibly US-brokered peace talks as we will discuss later in this book.

These events were devastating to the Palestinian aspiration to establish their own independent state. The U.S., Britain, and French were keen to consecrate the military superiority of Israel in the region and clandestine transfer of the French nuclear technology in the 1950s made it possible for Israel to develop its own nuclear arsenal

(Khalidi 2013). This U.S. desire of ensuring qualitative Israel's military superiority in the region has been a consistent policy until the present day. Nothing describes this policy better than the late Malcolm Kerr who contends:

> The pre-1973 record of American initiatives… indicates a pattern of too little too late, of grossly inadequate political support from the White House, and of a curiously persistent misconception that America must bring together Arab and Israeli governments that really want peace and successful negotiations, rather than that America should crack their heads together. Intended or not, the consistent effect has been to buy time in behalf of the status quo, which is to say, in behalf of the Israeli accumulation of *faits accomplis* and the Arab accumulation of resentment. (Kerr 1975)

Driven by its ever insatiable appetite for territorial appropriation, Israel launched another surprise attack (what is widely known as the Six-Day-War) on its surrounding Arab countries, namely Egypt and Syria, in 1967. In this war, Israel expanded its territories by annexing the rest of historic Palestine, as shown in Fig. 7. In six days Israel was able to gain control of the Gaza Strip, Sinai, the West Bank and the Syrian Golan Heights. This also brought death, destruction and displacement for the second time to the Palestinians, some of whom were already refugees and have not even recovered from the impact of the 1948 War. What ensued over the next fifty years was the implementation of the Zionist vision across the Palestinian territories with increased Israeli settlements expansion on private and public Palestinian land. At the expense of the Palestinians, water was diverted to benefit these Jewish settlers who also use a vast network of roads exclusively constructed to facilitate their travel while blocking the Palestinians from using them. These territories including the Gaza Strip, West Bank and the Golan Heights are not part of Israel and as such are considered and widely referred to as occupied territories. Akin to the 1948 *Nakba*, or "catastrophe," the Palestinians often refer to the Six-Day War of 1967 as the *Naksa*, or "setback," both of which represent pivotal junctures in the historical trajectory of Palestine (Munayyer 2017).

This war was manipulated and portrayed by Israel in its public discourse as a result of Arab countries' aggression against the small and newly established State of Israel which tried to defend itself, rather than the result of the more powerful, well-equipped, US-funded Israel engaging in a planned land grab to expand its territories. Israel began to introduce new draconian measures, such as special ID cards for the Palestinians (*Hawia*), closures, roadblocks, curfews, arrests and prolonged detention, house raids and demolitions, and brutal beatings, among other, to suppress and subjugate the Palestinian population fell under its control, many of whom are refugees. Palestinians living in these territories are under Israeli occupation and have no civil rights in the eye of Israel and are discriminated against by its modus operandi implemented policies and practices. Israel began to immediately beef up its security and build new illegal settlements inside of the 1967 land including the Gaza Strip and the West Bank in defiance of international law that prohibits Israel from doing so. Although these settlements were declared as illegal under the UN and the International Court of Justice, Israel continued to aggressively fund and expand them. Today, there are over 750,000 illegal Israeli settlers in the West Bank. These settlements are connected by a network of roads that are exclusively for the use of

Fig. 7 A map of Palestine in the wake of the Six-Day War, 1967. *Source* Illustration rendered by Michael N. Erickson, 2018 (adapted from Visualizing Palestine)

Jewish settlers on which Palestinians are not allowed to travel. As such, Palestinians cannot use these roads and their communities are effectively confined and separated from each other and from their farmlands (which are the main sources of livelihood for many Palestinians). In addition, these Israeli settlements are located strategically to ensure control over best farmland and key water resources, such as the Western Aquifer in the Palestinian West Bank, which are expropriated for the exclusive use of Israel's population. In 1979 Israel returned the Sinai to Egypt as part of a peace deal, but still maintains control of the Golan Heights until this day.

The Six-Day-War of 1967 was another critical juncture in the conflict because it severely altered the landscape of the region and access to the regional water resources, including ground and surface water resources. The new configuration and consequent positioning of the co-riparians in relationship to regional water resources also meant a vastly different allocation and utilization of these resources. Israel was caught in *flagrante delicto* with its expansive policies and territorial appetite that led to the events of 1948 and again the Six-Day War of 1967, and pari passu with these events is its ongoing practice of house demolition in the Gaza Strip and the construction of the illegal settlements and deliberate policies of segregation against the Palestinians.

Once again Israel reassured its military superiority and established itself as a hegemonic power in the region as the outcome of this war was detrimental to the Arab countries that lost part of their territories. In the wake of this war, thousands of Palestinian refugees were once again displaced into the neighboring Arab states as a result of Israel taking over the Palestinian Gaza Strip and the West Bank. Except for Sinai, which was returned to Egypt through the 1979 Israeli-Egyptian Peace Agreement, the new border and territorial configuration that the 1967 War created largely remain intact today. In addition to changing the international borders and the demographic characteristics of the region, this war also imposed a new reality pertaining to the co-riparians' new relationships as well as access to water. In this regard, these territorial changes also directly and significantly influenced access to water resources as an extension to the land. Israel emerged as the winner by securing full control of the Jordan River, Banyas River, and Lake Tiberias and further expanded its access to the Yarmouk River. This allowed Israel to utilize Lake Tiberias for its exclusive storage needs and prevented the West Bank Palestinians and Lebanese from accessing the River basin altogether. In addition, as a result of its control of the West Bank, Israel secured uninterrupted access to the Palestinian Mountain Aquifer, which remains key in its water security. The war also created a great deal of hydrological interdependency that led to heightened tension between Israel and the surrounding co-riparians. Not only was Israel able to utilize Palestinian surface and ground water, but it is also able now to prevent the Palestinians from adequately meeting their basic water needs. Similarly, through military means taking advantage of its military superiority, Israel was able to prevent Jordan from constructing the Maqarin dam to divert water from the Yarmouk River as stipulated by the Unified (Johnston) Plan, which forced Jordan to limit its water utilization to only the King Abdallah Canal and the water of the side wadis. Moreover, Israel has been extracting and transferring greater amounts of water outside the river basin, as far to the south as the Negev Desert, via its National Water Carrier. Collectively, this gave Israel a significantly advantageous hydro-political position compared to its neighboring countries that are barely able to meet their basic water needs.

Throughout history, Israel failed to abide by the many UN Security Council resolutions and instead created an unbearable situation for the Palestinians that under Israel's occupation are ipso facto treated as second class citizens. The mounting pressure and frustration that over twenty long years of brutal Israeli military occupation caused finally gave rise to the first Palestinian *Intifada* (uprising or shaking off) in 1987. The *Intifada* was a large-scale popular mobilization of Palestinian communities in the Gaza Strip and the West Bank to express their rejection of the continued occupation and demand their rights, freedom, and self-determination. It was not long after the eruption of the *Intifada* that the leader of the PLO, Yassir Arafat, was able to strike a deal with Israel in 1993 part of the Oslo agreement and what will be known as the peace process. Under this deal the Palestinians were allowed to establish a government and role themselves in the Gaza Strip and Jericho in the West Bank. As we will see in the subsequent chapters that examine the so called "peace process" and the ensuing events in greater details, the outcome of these talks was not short of a political debacle to say the least. It reinforced the Israeli occupation and failed to establish a Palestinian state, while allowing Israel to build more settlements on Palestinian land. Now, the Palestinian communities are further away from achieving peace and independence as a result of these so called "peace talks." Israel utilized these years of peace negotiation to concoct an insidious plan to prolong occupation and strengthen settlement while tightening its grip on the occupied territories and its natural resources including water.

The institutionalization of different policies regarding Israelis and Palestinians inside Israel personifies de jure discriminatory practice that is widely accepted and enforced by Israel. These de jure policies are enforced in a de facto fashion in the

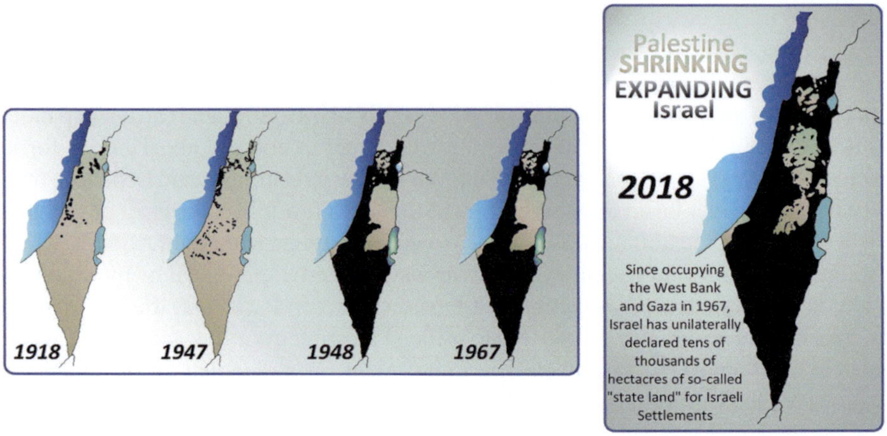

Fig. 8 The overtime cartographic regression of Palestine and the ever expanding Israel, representing the current situation. *Source* Illustration rendered by Michael N. Erickson, 2018 (adapted from Visualizing Palestine)

Occupied Palestinian Territories with much more aggressive policies whose aim is to confine and segregate the Palestinian villages in the West Bank and impose illegal siege on the entire population of the Gaza Strip. These policies in times of "peace" are often translated into a collective punishment as a matter of common practice and frequently escalate to outright wars in Gaza turning it into the world's largest open-air prison.

The result of the systematic expansion of land that characterizes Israel's policies since its inception can be clearly seen in how the map of Palestine and Israel has been progressively changing throughout the past century with an ever-shrinking Palestinian land and expanding Israel. Nothing can truly depict the magnitude of this reality better than the cartographic regression shown in Fig. 8. It provides irrefutable evidence not only of this systematic expansive policy whose main goal is to acquire as much land as possible with as little Palestinian population as possible, but it also provides yet another ironclad rebuttal of the false narrative about Palestine being as a land without a people for a people without a land.

Palestinians residing in the Occupied Palestinian Territories and abroad started to realize after over twenty years of futile negotiations that the peace process itself has been leveraged by Israel as a cover to impose harder restrictions on them, booster their settlements, increased assassinations, and implement a tighter siege on Gaza. During these years, a considerable amount of Palestinian land has been annexed by Israel for its illegal settlements and caused major land loss in the West Bank which became a collection of physically disconnected and disjointed Palestinian communities. The division between the Palestinian leadership in the West Bank and Gaza is a self-inflected wound that made things even worst for the average Palestinian on the ground. Due to its consequential impact on the overall geopolitical landscape, this division will be given special attention in Chap. 3 of this book.

Moreover, the loathing of an odious apartheid that is created by Israel is strongly felt throughout the Palestinian communities who view Israel as an occupying force. This feeling is indicative of major disappointment in the current peace process and what it produced, which is in turn indicative of the need for a practical solution that ensures the right of Palestinians to statehood and self-determination. Although significant aspects of the Arab-Israeli conflict were resolved between Egypt and Israel in 1979 and between Jordan and Israel in 1994, the Palestinian issue has yet to be resolved albeit decades of futile peace talks. Most Palestinians today view the "security coordination" that was established between Israel and the PLO as a result of these aforementioned peace talks as a major sellout of their rights and overall cause. It is one of the most hated aspect of the geopolitical relationship that was formed and reinforced by Israel and the United States. It became a widely recognized symbol of treachery and deceit due to what it meant for people on the ground many of whom are assassinated and executed as a result of this coordination and the relentless military activities led by Israel and supported by the PLO. The unilateral Israel's withdrawal from the Gaza Strip in 2005 and the election of Hamas trigged armed clashes between Hamas and its rival Fateh that resulted in Hams taking over the Gaza Strip. Since then Israel and Egypt implemented a siege on Gaza in 2007 that is in effect until this day, confiding nearly two million Palestinians in the world's largest and most densely populated open-air person depriving them from their very basic human rights. The rift in the Palestinian house, namely between the governments in the Gaza Strip and the West Bank, significantly impacted the overall unity of the Palestinians and will be discussed in details later in the book.

As military activities increased so did the siege shocking the Gaza Strip economy. Israel's policy in dealing with the Gaza Strip has taken a major shift in recent years and took the form of military invasions and armed conflicts. To date, there have been three major wars that Israel launched on the helpless and defenseless civilians of Gaza in 2008/2009, 2012 and most recently the 51-day war in the summer of 2014. These wars were carefully orchestrated in terms of both the military and the media campaign aspects to provide cover and justification for the Israeli onslaught.

What's noticeable in these three wars is the Palestinian resilience in the face of the cheer number of civilian casualties and the intensity and duration of the onslaught that continued for 51 days on the besieged population of Gaza, which is considered the longest war in the history of the Arab-Israeli conflict. There is a crucial difference between these wars waged against the besieged Gaza and the other wars that happened in the past in Lebanon or Syria—that unlike Gaza, civilian population is usually given the opportunity to leave and flee the areas where fighting erupts. In Gaza, however, people are besieged and trapped inside the largest open-air prison with nowhere to go even during brutal wars. The 2014 assault was documented in great details by Max Blumenthal in his book, *The 51 Day War: Ruin and Resistance in Gaza*, in which he describes the daily lives of the local Gazans and how they coped with the war. During this war, Israel launched air strikes on Hamas-controlled Gaza, followed by a ground invasion, after which Israel was forced to sign a ceasefire that has been in place since then, notwithstanding its lack of implementation given that the lifting of the blockade and opening the borders of Gaza which was one of its conditions has not been accomplished (Blumenthal 2015). Israel deployed its military might to deliberately target homes, hospitals and schools, in what has been described by many international and human rights organizations as an egregious war crime. In this war, Palestinians utilized attack tunnels in such a large scale operation to take the battle behind enemy lines from the land, air and sea. Israel had several goals, namely the demolishing of these tunnels, putting an end to the rocket attacks and dismantling the Hamas infrastructure. However, none of these stated goals, which have been constantly changing throughout the war depending on what Israel thought to be achievable, were accomplished. Instead, Israel killed well over two thousand civilians and injured and displaced thousands others, including women and children, causing massive destruction of most of Gaza's infrastructure including government buildings. These pretexts Israel used were proven to be strictly part of its propaganda and public discourse campaign to whitewash its ghastly crimes it committed in Gaza during the 51-day assault. As it failed to accomplish its stated goals, Israel (despite its military superiority) emerged unsuccessful both on the military and political fronts to say the least. The myth of the undefeated army that Israel has propagated throughout the region was dashed at the gates of Gaza. From the Palestinian point of view, although brought death and destruction, this situation has provided the Palestinians the impetus to bond the Palestinian communities.

The consequential result of these wars in terms of the high civilian death tolls once again represents yet another chapter of the Palestinian *Nakba* (catastrophe) which seems to have been propagated and exacerbated with the outright military operations and well-documented war crimes Israel committed in Gaza during all these three wars. Thousands were killed and injured and hundreds of thousands, who already has been displaced in 1948 and 1967, became once again homeless in Gaza.

Commenting on the high numbers of civilian causalities in Gaza, Noura Erakat, Palestinian-American human rights attorney, explains that "these numbers indicate Israel's egregious violations of humanitarian law, ones that amount to war crimes. Beyond the body count and reference to law, which is a product of power, the question to ask is, What is Israel's end goal?…. Israel cannot bomb Palestinians into submission, and it certainly cannot bomb them into peace" (Erakat 2014).

In the wake of these atrocities, the Palestinians became reluctant and increasingly critical of their current leadership's ability to bring peace and prosperity to their communities, let alone engendering self-determination. As such, both the PLO and Fateh in the West Bank and Hamas in Gaza lost a great deal of support from their local communities who got fed up with the ongoing violence for which they pay a heavy price. People now are looking for a way out and while it is difficult to imagine a peaceful end to this conflict, achieving peace is not impossible with political will from both sides.

The impact of these wars on the prospects of peace between the two sides cannot be trivialized as they acted as deterrent to bringing both nations closer to achieving peace. At the present time, prospects of peace seems to be nil with both Palestinians and Israelis losing hope in the peace process, and most importantly trust in one another as partners in any future peace settlement. In fact, the impact of these wars can still be felt so profoundly throughout the Gaza Strip as they turned an already dire situation into disastrous conditions, with limited water and power supplies. The water sector suffered the most as 90% of the water in Gaza is unfit for human consumption. According to the UN, Gaza is expected to be unlivable by the year 2020 as conditions continue to deteriorate and the infrastructure has been experiencing a process of de-development. As a result of the aforementioned unlivable situation that these peace negotiations yielded for the Palestinians, a new wave of frustration mounted into what has been described as a third Palestinian *Intifada* shaking off in the occupied territories, namely in the Gaza Strip with the continued Great March of Return demonstrations along the fence with Israel. The impact of these wars and the new wave of unrest will be further discussed later in this book.

The unorthodox United States election resulting in the election of Donald Trump as the U.S. president, the rise of ISIS in Syria and the involvement of Russia and the U.S. in this proxy war have taken the focus away from the Palestinian-Israeli peace process, which has been stagnant or largely nonexistent. In an unprecedented step on December 23, 2017, the UN Security Council voted in favor of a resolution demanding Israel to stop building settlements on Palestinian land. With the U.S. abstaining from vetoing the resolution, it allowed it to pass resoundingly creating widespread virulence and sending shockwaves into Israel who viewed the U.S. position as an abandonment of its legacy of vetoing such resolutions. Although Egypt withdrew it under pressure from Israel and U.S. president-elect Donald Trump who also urged the U.S. to veto it, the resolution condemned the construction of illegal Israeli settlements in the Palestinian territories describing them as a "flagrant violation" of international law and that these settlements have "no legal validity." Furthermore, it demanded the halt of "all Israeli settlement activities" in the Palestinian territories occupied in 1967, including East Jerusalem, describing it as an essential prerequisite for salvaging the two-state solution. Finding itself in a quagmire and flying in the face of international criticism, Israel seethed over the UN move and the role of the U.S. in it, calling on the new U.S. administration to repeal it, putting it on a collision course with the UN. In the words of its Prime Minister, Benjamin Netanyahu, Israel was quick to criticize and condemn the UN and flatly reject the new resolution describing it as "shameful and absurd," slamming the then current Obama administration for "abandoning Israel" at the UN and further blamed the U.S. of orchestrating the whole thing—an accusation that the U.S. vehemently denied describing it as absurd and groundless. In the wake of this decision and the diplomatic mayhem that ensued, Israel fears international pressure, including boycotts, sanctions and claims at the international criminal court that could ensue as a result to force Israel to abide by this resolution. By not using its veto to block this symbolic decision, the U.S. was criticized by Israel on the basis of its unusual action (or lack thereof) that signified a sharp rebuke to Israel's current policies in the occupied territories in relationship to settlement building. This, however, did not strop Israel from defiantly announcing its plans to increase its settlement building activities in the West Bank and East Jerusalem, which are areas that would potentially be part of the future Palestinian state, rendering the two-state solution impossible.

The Palestinians criticized Israel's reaction describing it as irrational. The Palestinians feel that such resolutions help in internationalizing their cause and gaining leverage to potentially force Israel to abide by international law, namely that the direct negotiation route did not yield any positive outcomes, and instead gave Israel cover to expand these settlements during the past twenty years of peace negotiations. Regarding this historic decision chief Palestinian negotiator, Saeb Erekat, haling it as a major step towards peace and reflecting the overall Palestinian stance, said "This is a day of victory for international law, a victory for civilized language and negotiation and a total rejection of extremist forces in Israel" (Aljazeera 2016). Erekat added "The international community has told the people of Israel that the way to security and peace is not going to be done through occupation… but rather through peace, ending the occupation and establishing a Palestinian state to live side by side with the state of Israel on the 1967 line" (Aljazeera 2016). One should put this decision in its larger context in relation to its timing and the conflicting positions between the departing and upcoming administrations and how this decision may impact the overall peace process. The bottom line is that this decision came as an expected outcome of Israel's ongoing expansive policies that need to be curtailed to prevent more physical changes on the ground and the associated damage to the prospects of a Palestinian state and instead allow for the peace process between the two sides to take its course. This resolution, however, will not be worth the paper it is written on if it is not accompanied by a solid mechanism that ensures its implementation, and *pari passu* with this, a renewed effort towards resurrecting the peace talks to provide an alternative strategy to the current negotiation deadlock. While it simply condemned the future construction of Israeli settlements, calling for its halt, the decision did not address the issue of dismantling the current illegal settlements that are already constructed in the occupied territories and akin to many UN resolutions failed to provide any practical guidance for its enforcement. There are many UN resolutions that Israel has yet to bide by due in part to the fact that these resolution lack real teeth.

It is also worth mentioning that, while the Palestinians are lauding about its impact on preserving the two-state solution, the decision itself does not necessarily provide remedies towards ensuring a specific final solution such as the two-state solution, nor does it exclude the one-state solution from future considerations. All it did is to pave the way for the removal of a major obstacle in the way of achieving peace between the Palestinians and Israel, and personified a fulfilment of a major prerequisite to future peace negotiations between the two sides, if Israel actually decides to abide by it. It remains to be seen what this actually means for both sides under the new Trump administration. Trump's announcement to recognize Jerusalem as the capital of Israel and the U.S. plan to move its embassy from Tel Aviv to Jerusalem marking the seventieth anniversary of the *Nakba*, sparked resentments and clashes. The impact of Trump's deal of the century remains to be seen and adds layers of uncertainty to the obfuscation that currently surrounds the Palestinian-Israeli relationship. The overall peace process, its shortfalls, prospects, and options will be discussed in more details in the following chapters.

Chapter 2
De-mystifying the Conflict: A Diagnostic Account of the Middle East Geopolitical Landscape

The tragedy in Palestine is not just a local one; it is a tragedy for the world, because it is an injustice that is a menace to the world's peace.

—Arnold J. Toynbee 1968

Representations of hydrological resources reflect competing geopolitical strategies whenever resources are potentially available from either side of a political border. Widely known as the mother of all conflicts, the Palestinian-Israeli conflict is regarded as the sine qua non for many disputes in different parts of the Middle East and beyond. The sustenance of this intractable conflict over several decades gave rise to many regional and global disputes and tension that can be evaded by eliminating the root causes of the injustice and frustration from which they emanate. With this in mind, this chapter will reflect on the way in which conflicting representations of hydrological resources have created tension, injustice, and conflict in the Middle East, with particular emphasis on the Israeli-Palestinian conflict, but also touching on the Golan Heights and other relevant issues.

As the first chapter provided a brief foundation of the historical context of the conflict in general essential for the overall understanding of the issues at hand, this chapter will provide a diagnostic account of the overall Arab-Israeli conflict in relation to water resources allocation mechanisms and the intricacy and dynamics of these aspects, emphasizing the role of water as a driving factor. These include political, environmental and physiological aspects, power structure imbalance, water and national security, water access and infrastructure and its impact on current water consumption, and perception of "the other," among other factors. Due to its profound nature and severity driving many aspects of the conflict in the Middle East, transboundary water resources, often considered a low-politics issue, is intertwined with issues of high politics and cannot be discussed in separation of the overall conflict. By tracing the nature of the consequential water crisis in the Middle East with its root causes and ramifications, this chapter seeks to provide evidence of lack of equitable water allocation in the current configuration and arrangement of the status quo in relation to water access and consumption patterns. This also opens the door

A. Abukhater, *Palestine - Peace by Piece*,
https://doi.org/10.1007/978-3-030-03985-1_2

for a new thinking that, while views equity as a prerequisite to achieving a mutually acceptable settlement, also provides ample justification for the need to apply equitable principles to promote cooperation and regional peace.

The concepts and analytical framework provided in this part represent universal principles that, while apply to the Middle East water conflict, are also reflective of, and applicable to many other disputes over natural resources around the world. This aspect about the research is particularly of great interest to the quest and scope of many other researchers. Unlike most water related literature, which strictly focuses on water alone with cursory attention to the relationship between water and other more pressing political and diplomatic links, this chapter will provide that missing link between low and high-political issues including water, land, sovereignty, and diplomatic relationships.

1 Conflicts Over Water Resources—Understanding Its Complexity and Realizing Its Potential

It is very unusual to find a shared water resource that has not witnessed some sort of conflictive relationships among those who share its water. History shows us that nations can clash over water in relation to both essential human and economic development needs. As the population of co-riparian states continues to grow, so does the demand for additional water, yet water is limited in nature, more so in arid regions. This situation is often accompanied with a high degree of complexity not only with respect to the relationship it creates but also the ongoing tension related to water allocation which can be always questioned and often challenged on a variety of grounds at any given time. This aspect about the complexity of shared natural resources is of great interest to the quest of this book. It is crucial to note that while water can play a pivotal role in alleviating political conflicts and hostility, water conflicts themselves are often nuanced in nature. This complexity can be attributed to the very nature of water, which tends to ignore political boundaries, flowing through various geopolitical, social, and economic divides (Abukhater 2013). Another factor that contributes to the complexity of water conflicts is the fact that the availability of water is always in flux making it difficult (if not managed well) to equitably allocate with a formula that can account for the variation in water availability throughout the different seasons of the year. This particular aspect is key not only in reaching agreement on water allocation plans but also in determining how successful and sustainable their implantation is. One may also think of water as an extension of land and property rights. As such, many view it as a matter of pure ownership solely related to property rights. Others may think of it as a matter of historical rights, reflecting previous usage throughout the years that need to be respected and accounted for. Others may also think of it as a matter of national security given its importance not only in domestic uses but also in industrial and agricultural production and the overall economic viability, nation building and security concerns.

All of these aspects and points of view about water reflect and explain the fact that humans often tend to compete for their share of water (in terms of its quality and quantity) for survival, consumptive uses, and economic development (Abukhater 2013). This complexity becomes much more pronounced when these aspects are conjoined with the natural, environmental, ecological, physical, psychological, spatial, temporal, technical, institutional, jurisdictional, and political aspects, among many others. These problems usually have multiple dimensions that slice through layers of uncertain variables related to the social, natural and political systems that are constantly changing across various scales and levels (e.g., local, regional, national, global). These are all ever changing, unpredictable and for the most part uncontrollable (Islam and Susskind 2013). As a result, leading scholars in the field attest to the complexity of water issues that are considered "wicked" problems. These wicked and multi-layered problems cannot be simply resolved based on a solely technological or scientific approach, but rather many other less objective measures and adaptive solutions that can only be achieved through negotiated settlements where key players and stakeholders can reach win-win situations that they can amicably endorse and successfully implement.

It is precisely this complexity that often ipso facto drives either conflicts or cooperation depending on the degree of diplomatic, political and strategic relationships among involved co-riparians. These relationships and configuration are all contingent upon the overall political, historical, and socioeconomic atmosphere of the shared river basin in general, as well as the individual and collective interests of riparian states sharing it. There are many facts and anecdotal examples from around the world that can clearly illustrate the complexity and the potentially conflictive nature of water when the latter is not properly managed (Allouche 2005). One such intriguing fact is the term "rivals" which is derived from the Latin word rivalis, which refers to two or more people who share the same river or stream (Dellapenna 2002; McDermott 2009). Due to its complexity and importance in either driving conflicts or cementing peaceful relationships, there are many examples where water emerged as a key determinant that constituted either an obstacle to achieving peace (such as the Israeli-Syrian relationship over the Yarmouk River) or a reason for achieving lasting and robust cooperative arrangements (such as the 1960 Indus Waters Treaty between India and Pakistan and the Lesotho Highlands Water Project (LHWP) Treaty between Lesotho and South Africa). It is very critical, therefore, to understand and leverage the great potential of water in making and sustaining peaceful arrangement among disputing nations. Water, due to its complex, multi-layered nature that defies geographic and political divides can potentially act as an incentive to achieving peace, quelling tensions, building trust and confidence, and fostering mutual social and economic development goals leading to a more amiable relationships among co-riparians.

2 A Framework for Understanding the Volatile Geopolitics of the Middle East

Being a semi-arid region, the Middle East is clearly an area where water scarcity constitutes a real urgent and pressing problem with consequences that far exceed its boundaries. Both the geopolitical landscape as well as the inhabitants of this region are equally impacted by water scarcity issues that drive many aspects of their very lives and perceptions. This cannot however be used as an argument to justify situations where entire communities or nations are deprived from accessing water for their basic needs, such as the Palestinian communities where water is often plentiful depending on their geographic location. The relationship between the hydrological resource use and geopolitical representations has always been intimately linked whenever these resources are potentially available from either side of a political border. Working in unison together, water use and geopolitical representations provide us with revealing diagnostic tools that help trace the roots of one of the world's most troubling conflicts involving water. As the first chapter demonstrated, while the Palestinian-Israeli conflict is clearly driven by reasons other than the actual allocation of water resources, water itself plays an instrumental role with crucial geopolitical dimensions that tend to contribute to, and sometimes sustain the overall dispute. Generally speaking, the water shortage issue in the Middle East is a direct result of the limited water resources available in this region, the exponential population growth driven by both natural growth and deliberate immigration and its consequent spike in water demand, coupled with the regional competition over transboundary water resources that is fueled by the overall conflict over land and natural resources. There are four major dimensions that characterize the volatile geopolitics of the Middle East. These dimensions include:

1. The environmental dimension;
2. The hydro-hegemonic dimension;
3. The psychological environment and ideological dimension; and
4. The political and legal dimension.

Figure 1 illustrates a causal path of how these geopolitical factors relate to, and influence the overall regional stability resulting in tension, injustice, and severe hydrological conflict. The interactions among these key four dimensions can only be described as "volatile geopolitics" that characterizes the overall Middle Eastern geopolitical discourse.

It is no surprise then that regional instability and a talk-flight situation ipso facto dominate the geopolitical atmosphere of the Middle East, where regional acrimony as well as peace agreements are witnessed on an intermittent but frequent basis. This talk-flight nature that the conflict ended up with created a situation where both sides are perpetually responding to the actions of the other side, fueling a race to the bottom.

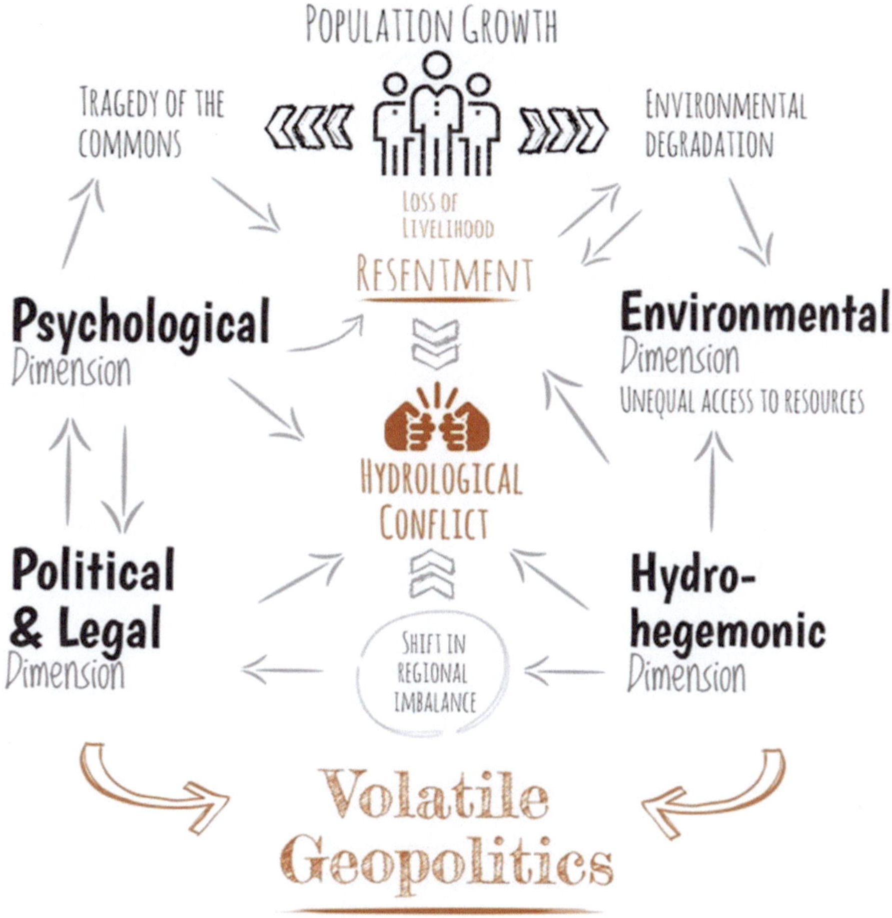

Fig. 1 Hydrological conflict and volatile geopolitics

2.1 The Environmental Dimension

2.1.1 Water Access, Infrastructure, and Variable Trends of Water Consumption

Understanding how access, or the ability to benefit from water, rather than rights, determines who can realize benefits from what resources at what time helps in understanding the overall situation in the Occupied Palestinian Territories. The spatial and temporal dynamics of military control primarily define who has access to common-pool resources and who gets excluded from accessing them. Although commonly understood as a function of gaining benefit, these dynamics of access are also a function of exclusion and prevention, clearly delineating winners and losers and leading in many cases to resentments and conflicts (Schwartz 2006).

This is evidently clear in the Palestinian-Israeli conflict. Palestinians and Israelis alike vie for their fair share of water with an unprecedented level of hostility. Not only have levels of this hydro-hostility become more potent as time wears on with no real solution in sight, but the nature and content of the discourse has changed as well. The larger political and cultural clash has emerged as a major contributor to the regional hydro-conflict, which reduces the chances of success of any attempt to craft a meaningful and practical resolution. Much of Israel's expansive policies and activities to annex into the Occupied Palestinian Territories, discussed in the first chapter, is driven or at least influenced by the desire to gain, control, and maintain access to water resources which evidently gets transferred into Israel for domestic, agricultural and industrial uses. As a result of this systematic expansion and inter-basin transfer accompanied by access restriction policies, water access is not equally shared between the Israelis and Palestinians. In the wake of the 1967 War, Israel secured access to major water resources in the West Bank and the Gaza Strip and prevented Palestinian communities from developing their own water infrastructure, making them water-dependent on Israel (Khalil 2016). Characterized by legal and technological imbalances, this situation illustrates a classical example of how access to shared natural resource is largely contingent upon power (Korf and Funfgeld 2005).

The problem partially lies with the advantageous hydro-political position that Israel holds, as a direct result of the 1967 War, which allows it exclusive control of over 85% of all available water resources in the region, leaving other riparian countries with less per capita water availability than that of Israel (Zeitoun 2005). As an upstream riparian, Israel diverts over 75% of the Jordan River's water before it reaches the Palestinian West Bank (PASSIA 2002). While Lebanon, Israel, Syria, and Jordan share the Jordan River, Palestinians are the only riparian with no access to the water of the river. Although the Johnston Plan of 1959 clearly constitutes their share of the Jordan River water, approximately 200 MCM per year (Rouyer 2000), Palestinians today are denied all access to the river to the east, in addition to most of the productive zones in the Western Aquifer Basin to the west, as shown in Fig. 2 (Zeitoun 2005).

On the technological front, Israel is far more advanced than the other co-riparians sharing these regional water resources. This superiority in technological advancement, which is used here as an indicator of economic conditions, gives Israel the upper hand by allowing it to extract and divert much larger amounts of water compared to other riparians. For example, this technological advantage can be clearly seen in the distribution and depth of the Israeli wells and their associated extraction capacity that Israel is able achieve which is shown in Fig. 3. The use of more sophisticated technology in water installations makes it easy for the Israelis to dig deeper wells. The spatial distribution of these wells and their ability to dig deeper wells compared to the shallow Palestinian wells results in a higher Israeli excavation capacity. These extraction activities from the deeper Israeli wells often cause the old and shallow Palestinian wells to dry up, which drastically affects the amounts and quality of water that the Palestinian communities receive. This technological prism

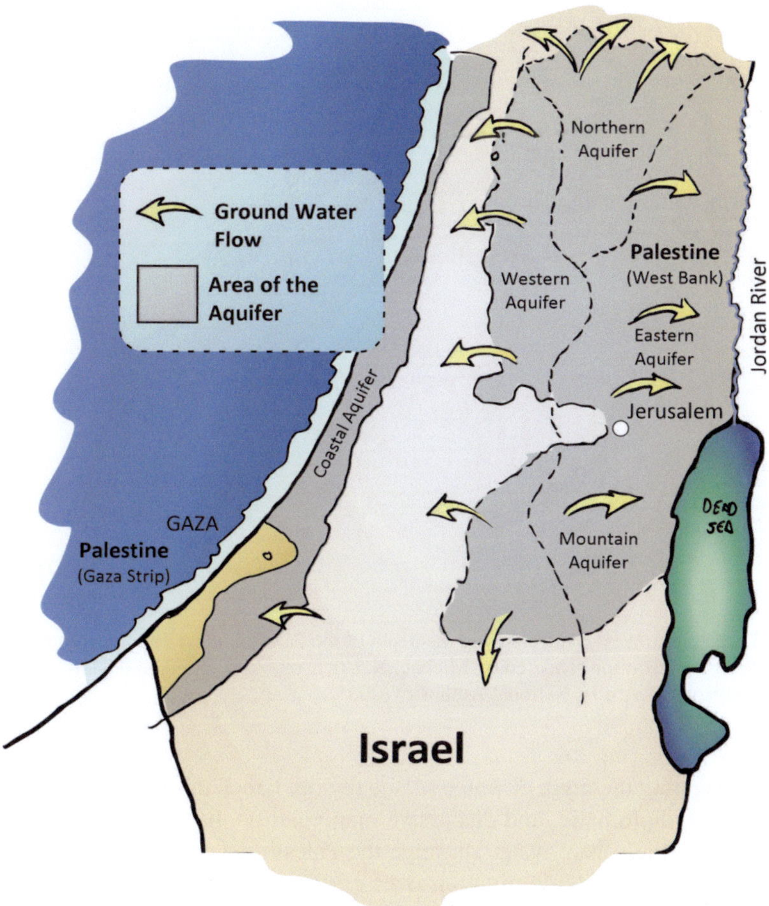

Fig. 2 Groundwater resources in the region. *Source* Illustration rendered by Michael N. Erickson, 2018 based on data from UNEP (2002)

connotes the extent of the unequal access to the same quality and quantity of water that the Israelis and Palestinians can actually extract.

In addition to its technological advantage, Israel also imposes strict regulations to impede the Palestinian communities and farmers from the ability to obtain their adequate share from these shared water resources. Part of these restrictions is a physical one—the Israeli wall that slices through the West Bank villages separating Palestinian communities from each other and from accessing water and productive agricultural zones. On one side of the wall, you can see lush Israeli communities with roads and services, while on the other side of the wall you can see the deprived, overpopulated Palestinian camps. Ironically, but not surprisingly, the path of this wall is almost entirely located inside the Palestinian West Bank.

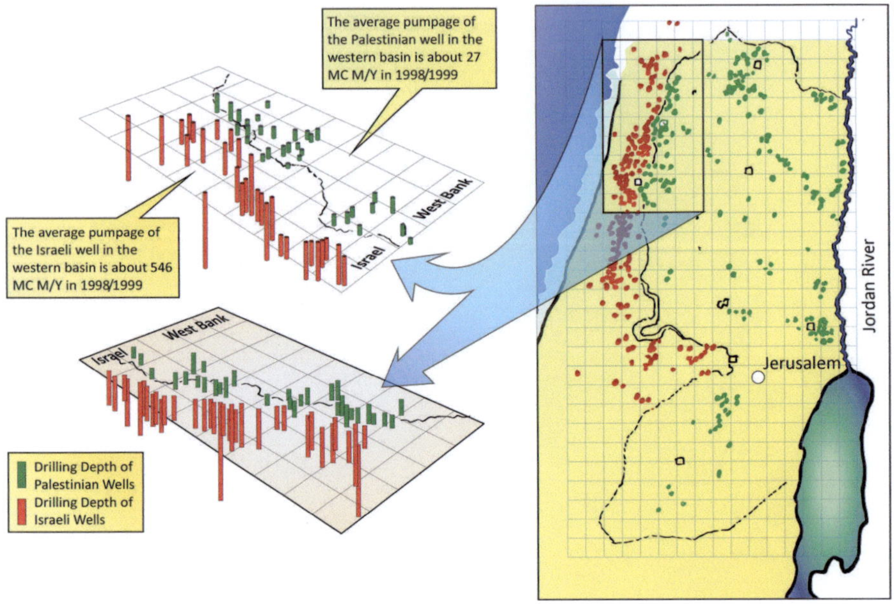

Fig. 3 A map of the West Bank showing the depth of the Palestinian wells compared to that of the Israelis. *Source* Illustration rendered by Michael N. Erickson, 2018 based on data from Palestinian Water Authority, Palestinian National Authority (2005)

By propagating the myth of water-scarce region, Israel, through a torrent of assiduous political, diplomatic, and discursive manipulation, has been successful in not only justifying the severe water shortage the Palestinian communities suffer from, but also in extending its domination over essential water resources of the region (as shown in Fig. 4). One only need to consider the fact that Ramallah receives more precipitation than London to realize the extent of this manipulative narrative, which exposes these untenable Israeli argument to deny the Palestinian communities their fair share of the regional water resources (Corradin 2016). Another restriction, which was instituted by the 1995 Oslo II interim agreement, requires any Palestinian project in the West Bank to seek approval from the Joint Water Committee (JWC) while giving Israel the de facto voting power (Khalil 2016). This caused many Palestinian water projects to be rejected, creating water access and extraction gap between Israel and Palestine. This means that the Israeli-imposed restrictions on the Palestinian communities in relation to accessing these resources, along with their pronounced technological disadvantage, results in their inability to reach deep levels of the aquifers. Consequently, this Israeli control over Palestinian water resources impeded the Palestinians from building their own water infrastructure causing the Palestinian communities to be largely dependent on Israel for their water needs, and as such, contributing to widening the water extraction and usage gap between the two nations, making it even more pronounced.

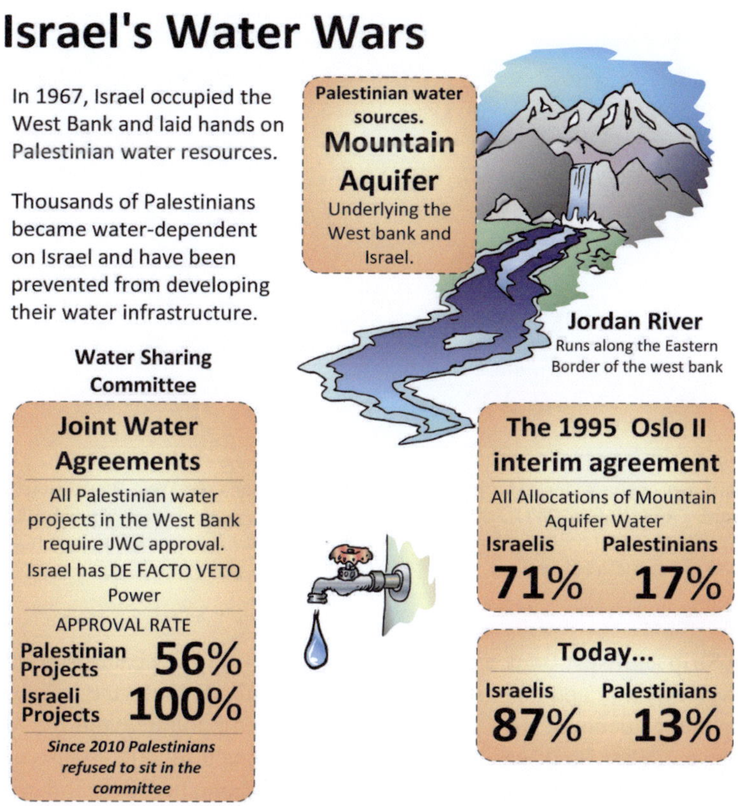

Fig. 4 Israel's water wars and access to water. *Source* Illustration rendered by Michael N. Erickson, 2018 based on data from United Nations office for the coordination of humanitarian affairs, cited in Corradin (2016)

Due in part to the above-mentioned factors concerning the Palestinian water access which is tightly controlled by Israel, and technological and infrastructure limitation, access to adequate water available to the Palestinian communities is restricted causing them to suffer severe shortages in terms of the quality and quantity of water to meet their needs. This created a situation that can be best described as inequitable in terms of accessing and allocating shared water resources. Even during normal consumption periods, the per capita water that the Palestinian communities in the West Bank for example receive is well below the World Health Organization minimum standard of 100 L per day (refer to Fig. 5).

Overall, it is estimated that the per capita Israeli water consumption is equivalent to more than five times the per capita Palestinian water use, when accounting for both the West Bank and Gaza Strip (Zeitoun 2005). Moreover, these comparatively low per capita Palestinian water use figures show a dramatic and alarming declining trend over time indicating an eminent water crisis, which is already taking place in many areas in

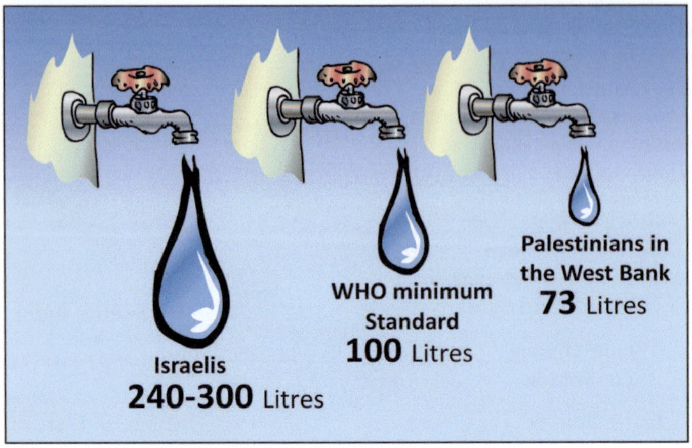

Fig. 5 Daily per capita water consumption patterns in Israel and the West Bank compared to the World Health Organization (WHO) minimum standard. *Source* Illustration rendered by Michael N. Erickson, 2018 based on data from United Nations office for the coordination of humanitarian affairs, cited in Corradin (2016)

the Gaza Strip and West Bank where water shortage is severely catastrophic in terms of both water quality and quantity. The Coastal Aquifer in Gaza, the main source of water for the residents of the Gaza Strip, is irreparably polluted and the Mountain Aquifer in the West Bank has been primarily dominated by Israel's use and inter-basin transfer. Demonstrating a stark contrast to the parched Palestinian villages, the situation is quite different in the nearby illegal Israeli settlements, where Israeli settlers enjoy ample water supplies with swimming pools and well irrigated gardens and farms (Amnesty International 2009). Because of these dramatic differences, attributed in part to the lack of Palestinian access to water along with the intermittent water supply that the Palestinian communities endure, water distribution between the Palestinians and Israelis reflects a situation of severe inequity, at best, and a "water war," as many scholars characterized it, at worst (Khalil 2016).

For the Palestinians, a people struggling for a subsistence level of water access, the situation deteriorates irrevocably as time goes by with no feasible resolution in sight. It is surely to the detriment of the whole region's future peace and prosperity that the Israelis have expropriated not only the Palestinians national and land rights, but also the right to have access to clean and reliable water supply. Many people in the Occupied Palestinian Territories are frustrated as a result of these intentional and widespread Israeli policies and often accuse Israel of using water as a weapon against them as collective punishment (Figs. 6 and 7).

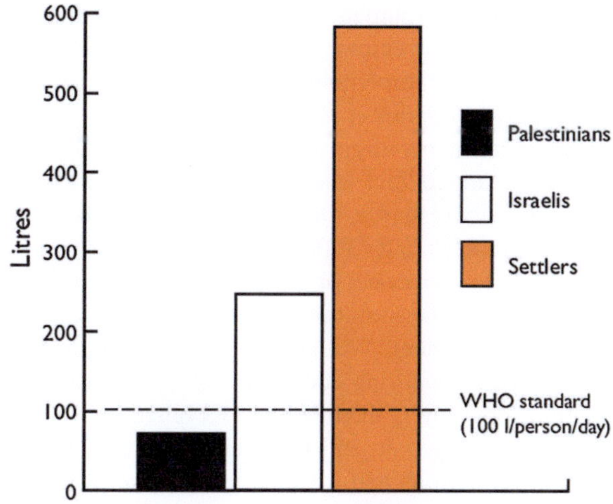

Fig. 6 Daily per capita water consumption. *Source* Based on data from PASSIA (2002); and FAO-AQUASTAT (2002)

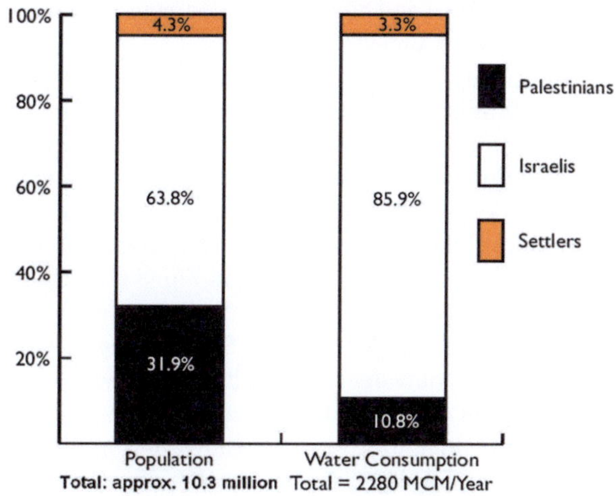

Fig. 7 Percentage population versus water. *Source* Based on data from PASSIA (2002); and FAO-AQUASTAT (2002)

2.1.2 Environmental Consequences

Population growth means elevated water demands, on one hand, and increased competition over water access, on the other. This increased competition and the availability of common-pool resources lead to a more consumptive and exploitive water use pattern responsible for the most part for the "tragedy of the commons" syndrome. In this situation each co-riparian tends to make rational decisions to divert the greatest amount of water and prevent others from doing the same. This creates a competitive, zero-sum situation and can easily lead to inequitable water allocation, which has a number of substantial environmental ramifications. Under this scenario, a win-win situation is hard to attain when there are clear winners and losers, including for example the environment which can be one of these losers that suffer as a result of this tragedy of the commons situation.

Within this competitive climate, hostility becomes more pronounced and the prospect of cooperation is severely crippled by the "prisoner's dilemma." The prisoner's dilemma paradigm contends that one of three situations may occur, including (1) if all states defect and choose to compete to maximize their benefit from the common-pool resource, the commons will be destroyed; (2) if one cooperates while the other refuses to do so, the "sucker's payoff" involves substantial losses to those who chose to cooperate; or (3) if all choose to cooperate, these significant losses will be avoided and the common-pool resource can be potentially protected through cooperative arrangement (Wolf 1995; Zawahri 2004). The behavior and strategy chosen by each state regarding cooperation or conflict depends mainly on the geopolitical relationships among them, which is a function of water availability, level of hostility, distribution of power, and legal and constitutional settings as well as religious and cultural affiliations and divides. In the case of the Middle East conflict, it is clear that the relationship between the two parties involved in the conflict is governed by hostility, mistrust and high level of competition that can lead to the depletion of these regional water sources.

The tragedy of the commons situation that exists in this region and its implications on the environment are both extreme. Environmental ramifications can be severe and range from the deterioration of the quantity and quality of water for the Palestinian communities to the salinity of the Jordan River and the declining trend of the Dead Sea water levels. The current excessive extraction of the Jordan River water and the consumptive patter of water use in this area caused the disappearance of the Dead Sea at a staggering rate, as shown in Fig. 8 (Greenberg 2006). The decreased inflow of the Jordan River as a result of this increased extraction increased its salinity and resulted in the demise of the Dead Sea, causing its water level to drop about one meter each year (the equivalent of 3 ft). It is expected that within the next three decades the Dead Sea will dry out and completely vanish, and will truly live up to its own name if nothing is done to save it. As shown in Fig. 8, the Dead Sea has been declining and shrinking for the past few decades, an unfortunate environmental consequence

Fig. 8 Satellite imagery of the Dead Sea in 1980, 2005, and the future. *Source* Illustration rendered by Michael N. Erickson, 2018 based on data from ABC News (2005)

that is expected to continue so long as the current excessive water extraction levels are maintained in the absence of environmentally friendly demand and supply-side management policies that take into account the health and integrity of the ecosystem.

Another environmental consequence can be found in the severe depletion and irrevocable contamination of the Coastal Aquifer in the Gaza Strip. It is estimated that more than 90% of the Gaza Strip's potential drinking water is not compatible with the standards of the World Health Organization (WHO) for human consumption due to this high level of contamination (Palestinian National Information Center 1999). This has serious and disastrous ramifications for the nearly two million human beings living in the Gaza Strip without other alternative water resources that can sufficiently meet their demand for clean and adequate water supply. This situation, which is exacerbated by the ongoing illegal siege, has been identified as a major impediment to any future peace negotiations, and further constitutes a major threat to the overall regional stability and security. Fair allocation of water resources is warranted in this situation to put an end to this tragedy. The difficulty however of attaining "fair division" of disputed resources becomes more severe when the parties claim sovereignty and entitlement, which is clearly the case in the Palestinian-Israeli conflict. Accomplishing this fair division, which means proportional distribution of costs and benefits among all parties involved, requires mutual agreement on specific criteria for what the parties consider fair and the selection of valid procedural rules to ensure its implementation.

2.2 *Hydro-Hegemonic Dimension*

2.2.1 Power Structure Imbalance

The shortage of adequate supplies of fresh water is considered as one of the most severe conundrums and pernicious issues that impede the Middle East peace process. In this context, as competition over shared water resources increases, decisions regarding water allocation have become increasingly contentious and controversial, highly politicized and tightly associated with the overall political discourse, potentially leading to disputes. As conflicts intensify, the interplay between controversial decisions regarding water acquisition and the degree of power-structure imbalance becomes more evident and pronounced.

Scholars of "theory of access" assert that those with greatest access to power are in the best position to control access to water resources (Peet and Watts 2004). In that sense, access to water is strongly linked to and influenced by the degree of hydro-hegemony that each co-riparian possesses. According to Zeitoun (2005), hydro-hegemony has multiple elements, including power, hydro-strategic position, and exploitation potential, all of which can directly or indirectly influence the outcome, analysis, and perception of any water access and allocation scheme. Based on this criteria (illustrated in Fig. 9), Israel is undoubtedly the most powerful state in the region, which partly explains its advantageous access to regional water resources,

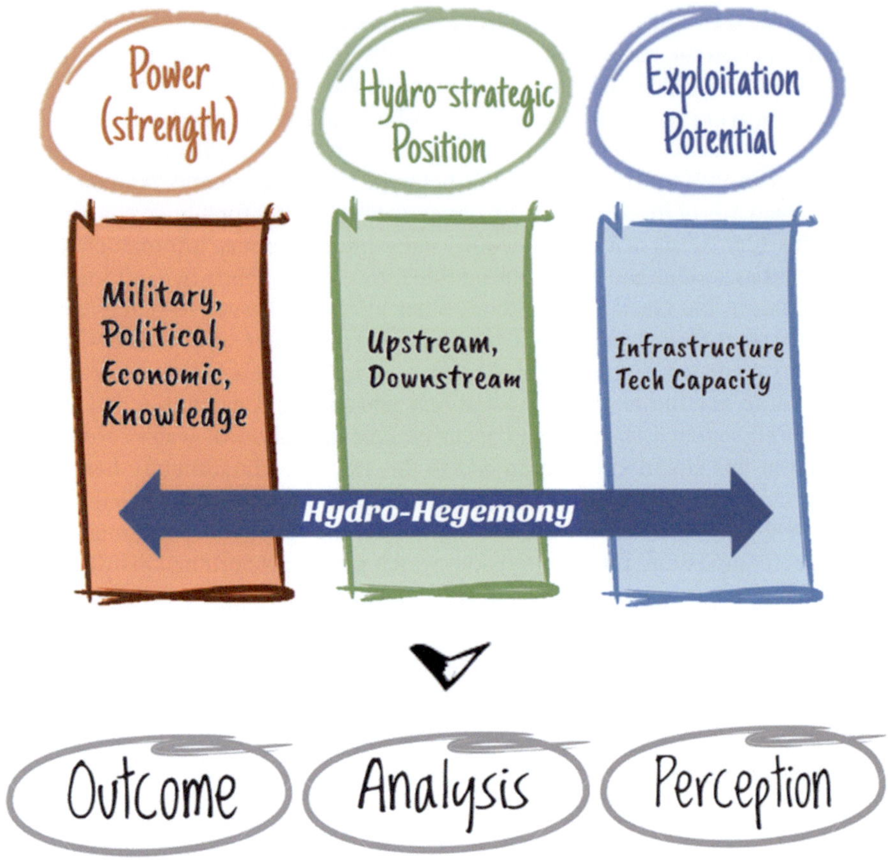

Fig. 9 Dimensions of hydro–hegemony

including both surface and groundwater resources. Realizing the significance of hydro-hegemony as a driving factor in determining who gets access to water and the quantity and quality of water they can extract clearly explains the significant volumes of water that Israel is able to secure. The interplay between water acquisition and hydro-hegemony, on one hand, and political decisions of either cooperation or altercation, on the other, is extremely evident in the overall water allocation policies in the Middle East in a number of ways. First, the riparian states' inability to facilitate stable cooperation has become an undisputable fact. The region has been engulfed with conflict and instability. As violence in the region gives rise to potential armed conflicts, the more powerful states tend to take advantage of the situation in order to maximize their share of water and undermine that of others who are seen as less powerful. For example, in the case of the Jordan River, many states, including Palestine, Israel, Jordan, Lebanon, and Syria, are riparians and have the right to share access to the river water. However, due to the political instability of the region, the only state that disproportionately benefits from accessing the Jordan River is Israel, given its economic and military advantages, which after all shapes many of its decisions in this regard. Conversely, there is ironclad evidence that the Palestinians, given their weak hydro-hegemonic position, are in a less favorable position, even when compared to the Jordanians, Israelis, Lebanese, and Syrians, as far as access to clean and reliable water resources is concerned. As such, these results are reflective of the dynamics of regional power imbalance, which can adequately explain the disproportionately inequitable water allocation in this part of the world.

Second, due to its perceived threat to its benefit as a result of cooperative agreements, the most powerful riparians are usually reluctant to entertain any option that would include cooperative arrangement to regulate their use of transboundary water resources. This is because cooperation is perceived by powerful riparians as a threat to its interests and can potentially jeopardize their most favorable position as far as water access and expropriation are concerned (Lowi 1995). Cooperation is only desired by powerful riparians in situations where it might be coercive, unsustainable, and sought to solidify and legitimize the status quo. By the same token, given the weak Palestinian political and military stance, most of the previous agreements achieved between the Palestinians and Israel fall into this category of coerced negotiation. Due to their evident disadvantage in the current water allocation schemes, the Palestinian negotiators were not in a position to reject any proposal by Israel and as a result they unwillingly acquiesced to plans that were not fair or just. Their negotiation strategy can only be characterized as "shooting at everything that moves," with little or no consideration given to the strategic value or practical benefit of these agreements, given their desperation for any resolution that could potentially provide a departure from the status quo. Any proposal was a viable one and was viewed as a better option than what they currently have in reality.

Another prominent example that shows the impact of this factor is the issue of the Syrian Golan Heights, which has been occupied by Israel since the Six-Day War of 1967. As shown in Fig. 10, due to its strategic location that overlooks northern Israel and southern Syria, the Golan Heights is considered by Israel as a vital piece of land that has significant strategic importance. As discussed in the first chapter, the Six-Day War of 1967 allowed Israel to gain de facto control of the Golan Heights which also meant that Israel now is able to secure access to Lake Tiberias, which Israel uses to store water and reportedly provides over 30% of Israel's water supply (Golan 2007). Syria, through military means in the 1973 Middle East War, tried to reclaim its control over the Golan Heights, but was unsuccessful because of Israel's strong hydro-hegemonic position which played a pivotal role in determining the outcome of the war and ensuing situation. One of the outcomes of the war was that Syria was forced to sign an armistice in 1974, based on which a UN observer force has been deployed to maintain a ceasefire (BBC News 2007). However, Israel's decision to unilaterally annex the Golan Heights in 1981 was a clear personification of the impact of its most powerful hydro-hegemonic position on the overall outcome and configuration of the geopolitical discourse in the region. As a result of its annexation of the territory, Israel intensified its settlements and applied its law and administration in the territory to change the reality on the ground and primitively prevent any

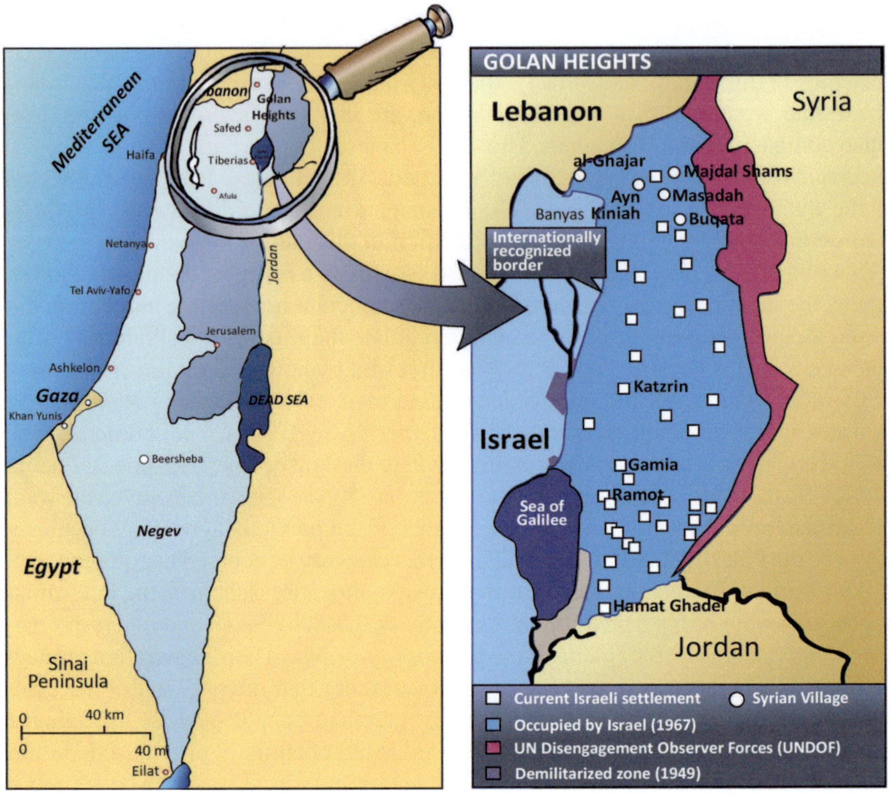

Fig. 10 Golan Heights' strategic location. *Source* Illustration rendered by Michael N. Erickson, 2018

future Syrian attempt to regain control of the territory. As is the case with the Israeli settlements in the Occupied Palestinian Territories, these settlements are also illegal and were not recognized by the international community. This situation has created tension and severely strained the relationship between the two countries until today.

Third, inequitable access to shared water resources can result in a state of resentment and hostility against the most powerful state because of its advantageous position as far as access to water is concerned. This hostility and resentment can potentially create political mayhem inevitably exacerbating international disputes and straining the relationship among different disputing nations. Hostility also gives rise to the unrealistic and overestimated water demand figures that can be misrepresented by each nation to justify their increasing water use, a situation that often undermines chances of reaching a sustainable settlement. This increased level of hostility can alter the very nature of the discourse by exacerbating the competition among riparian states to access more water and prevent others from doing the same. With the ever growing population of the region and the fixed regional water reserve, the demand

and therefore competition over water increases. Reflecting this geopolitical reality, hostility and competition between the Israelis and Palestinians to acquire more water is very high today and has provoked a situation of distrust and overall regional instability. The compatibility of the impact of hydro-hegemony and the distribution of water resources in shifting the regional balance and fueling and sustaining the Palestinian-Israeli conflict is unquestionable.

2.2.2 Water and Israel's National Security

As discussed in previous sections in this chapter, states require, and *rationally* vie for, their share of a continuous and adequate supply of food, water, fuel and other raw materials. The ecological security of states and nations is tightly linked to both national sovereignty and the need for a robust and secured statehood access to these resources for economic and nation building purposes. Securing this access to maintain ecological security means, and often necessitates, retaining exclusive control of one's land along with own natural resources linked to the land (Westing 1986). As a result, the perceived importance of land in relationship to sovereignty extends to also include natural resources. The strongly interconnected nature between national security and these elements including water and other natural resources can potentially create tension that may trigger or contribute to exacerbating conflict among different nations, all of whom will tend to strive to *rationally* acquire more of these shared and disputed natural resources. It is precisely this *rationality* in making decisions regarding access to land and natural resources that constitutes a significant part of the problem surrounding water security and access, namely when the question pertains to a common-pool resource.

This understanding of the dynamics between land and natural resources (including water) provides clear explanation for how Israeli policy makers often link water access to Israel's own national security. This is why Israel regards water as a strategic natural resource essential to its national security and political stability, on one hand, and further utilizes it as an effective weapon to put pressure on other Arab states, including the Palestinians, to direct negotiations for its benefit, on the other. Because of its strategic importance, the topic of peace and security is given high priority and introduced in the first chapter of the Likud Party Platform. Drawn upon this belief system that glorifies security and sovereignty, Zionists assert that "the right of the Jewish people to the Land of Israel is an eternal right, not subject to dispute, and includes the right to security and peace" (Virtual Jewish Library 1996). Although they acknowledged and emphasized the need to enter into negotiated agreements with the Arabs to reach peace, they further qualified this peace with security as a first and foremost condition stating that "security is the basis for durable peace in our region. Israel will make security a first condition in any peace agreement" (Virtual Jewish Library 1996). Israel's foreign policy is also strongly anchored around asserting and ensuring a high level of national security which is given great priority and emphasis in any diplomatic effort that Israel undertakes. To that end, the Likud Party Platform holds that "it will recognize the facts created on the ground by the various accords, and will act to reduce the dangers to the future and security of Israel resulting from these agreements" (Virtual Jewish Library 1996). It is no surprise then that Israel has been so adamant about its security concerns that in many cases these concerns trumped or undermined the chances of reaching peace with its neighbors. In the case of its negotiations with the Palestinians, these security concerns were emphasized as a prerequisite to these negotiations and were often used as an excuse to deny peaceful settlement proposals on the basis of lack of security.

Land and, by extension, water are both regarded as critical pillars for Israel's national security. Given Israel's control of most of the Palestinian natural resources in the West Bank, and its illegal settlements' strategic locations on the best catchment areas, it is estimated that about 70% of Israel's consumption of groundwater, and more than 33% of its overall annual water supply, comes directly from the Occupied Palestinian Territories (Lonergan and Brooks 1993). This high dependency of Israel on Palestinian water resources and the importance of water as a matter of national security for Israel are what motivated the former Israeli Agriculture Minister Rafael Eitan to declare that relinquishing Israel's control of the Palestinian water resources would "threaten the Jewish state" (Lonergan and Brooks 1993). This issue continues to be one of the most contentious ongoing hurdle to the current and future negotiations regarding the West Bank. Any plans that involve relinquishing these Palestinian areas will be challenged and potentially aborted on the basis of Israel's national security concerns. As such, previous negotiations failed to yield any results that would contradict the status quo Israeli control of these areas captured in the wake of the Six-Day War of 1967. This irreconcilable issue is also expected to hinder future rounds of negotiations if not addressed with a different mindset and approach that recognizes the need to end occupation of these Palestinian areas.

Israel's control of the Golan Heights, which represents a territorial and natural resources conflict, provides another example of how Israel's national security concerns shape its international relationships as well as the actions it takes. The value of the Golan Heights to Israel cannot be overstated and can be linked to the strategic importance of its location and the access to critical natural resources that it provides. The strategic importance of the Golan Heights stems from a number of factors. First, because of its central location between Israel and Syria and its significant elevation, the Golan Heights provides a natural buffer against any potential Syrian military invasion by land (BICOM 2007). The Israeli-Syrian relationship has not been particularly amicable since Israel secured control of the Syrian Golan Heights in 1967, coupled with the lack of a comprehensive peace agreement to settle the ongoing dispute between the two countries. Second, Israel will risk being potentially vulnerable to attacks that can be launched from the Golan Heights due to its high elevation if the area is not controlled by Israel. As a result, Israel has rejected, and is expected to foil any negotiations or proposal that would suggest relinquishing its control of the Heights. Third, because of its location on the east side of Lake Tiberias, the Heights provides important water storage capacity and a significant source of fresh water supply for Israel (about one third of Israel's annual water budget) (Golan 2007).

As such, the issue of the Golan Heights has severely strained an already troubled relationship between the two countries, which is made even more complicated with the current ongoing unrest in Syria as a result of the Arab Spring and the Russian military intervention in the country's civil war. Little has been done to address the issue of the Golan Heights which remains to be a stumbling block in the way of negotiation between Israel and Syria. During the 1999 talks, Syria demanded a full Israeli withdrawal to the pre-1967 border (Wihbey and Berman 2000). However, because of the fact that this Israeli withdrawal would give Syria control of the eastern half of the Jordan River headwaters and the eastern shore of Lake Tiberias, one of the main water resources and the main storage facility for Israel, Israel refused this condition for negotiation. If Israel decides to withdraw its presence from the Golan Heights, it would undoubtedly shift the hydrological balance of power in the region in favor of Syria, a compromise that Israel is not willing to make (Wihbey and Berman 2000). To that end, because of the Israeli government's view of the Golan as an essential asset to Israel's security and water needs, retaining Israel's sovereignty over the Golan Heights is identified as a major prerequisite for any negotiation with Syria. In addition, this negotiation may potentially yield a deal with Syria that compels Israel to dismantle its Jewish settlements in the region. In 2003, the Syrian President Bashar Al-Assad, reflecting the overall Syrian desire to regain access to the Golan Heights through negotiation channels, announced his country's intention to revive peace talks with Israel. A public opinion survey conducted in 2004 shows that the majority of Israelis diametrically oppose negotiations that involve returning the Golan Heights to Syria (BBC News 2007). It is understandable however that the internal unrest and popular demand triggered for the most part as a result of the Arab Spring for the Al-Assad's government to step down may have dwarfed the importance of the Golan Heights to the current regime in Syria, which sought help from Russia to quell the uprising and has committed horrific war crimes against its own people to remain in power. These gruesome war crimes and atrocities that continued for many years since the onset of the Arab Spring in 2011 have resulted in the murder of hundreds of thousands of helpless Syrian civilians and the displacement of millions triggering a wave of Syrian refugees into surrounding Arab countries and even Europe, which is known as the refugee crisis. The magnitude of the ongoing Syrian war, along with the rise of ISIS and the proxy war in Syria between the USA and Russia with the direct or indirect involvement of many regional players such as Iran, Israel, the Lebanese Hezbollah, Saudi Arabia, and others makes the issue of the Golan Heights secondary compared to these gargantuan issues facing Syria today as the nation's civil war drags on. In addition, the intensive and increasing presence of the Israeli settlements in the Golan Heights and the dependence of the State of Israel on water originated in this territory, is indicative of the Heights importance and suggests that Israel would continue to maintain control of the area, which is deemed as a matter of national security. Although the Al-Assad's regime has more pressing issues to deal with related to the ongoing internal unrest, reconciling these two radically different points of view about this particular area is proven to be an essential step towards achieving any progress in future negotiations. The complicated situation created by the Syrian war which is developing into a full-blown proxy war between the US and Russia, along with the strategic importance of the Golan Heights to Israel, indicates that these negotiations between the two countries over the Golan Heights will most likely not happen any time soon. Unless the current Syrian crisis is addressed as part of a comprehensive diplomatic effort that also takes into account the issue of the occupied Golan Heights, the status quo is likely to continue in the future and Israel emerges as the clear winner in this situation.

The land-water nexus entrenched in the Israeli ideological, political, and religious identity is best captured by the words of Haim Druckman, a well-known Israeli leader, who asserts that “settlements are the essence of our existence and flesh of our flesh. We shall not accept the amputation of our living flesh” (Lustick 1994). With this in mind, one must be cognizant of the implication of such attachment and commitment to settlements, which explains the current negotiation deadlock as a result of Israel’s unwillingness to negotiate or accept any agreement with the Palestinians or Syrians that may lead to compromising or dismantling these illegal settlements in the Occupied Palestinian Territories or the Golan Heights. Given its hydro-political importance, these Israeli settlements in the occupied territories appear to provide a safety net to ensure Israel’s national security and are perceived by Israel as an essential part of its very existence.

Water in these occupied territories is as important to Israel as land itself. According to the 1996 Likud Party Platform, “Jewish settlement, security areas, water resources, state land and road intersections in Judea, Samaria and the Gaza Strip shall remain under full Israeli control” (Virtual Jewish Library 1996). It also holds that “Israel will keep its vital water resources in Judea and Samaria. There shall be no infringement of Israel’s use of its water resources.” Water seems as important as Israel’s national security and almost a non-negotiable matter to the mainstream Israelis. This is a remarkable nuance that negotiators must be fully aware of. The persistence on the deliberate biblical use of “Judea and Samaria” to refer to the West Bank has a strong connotation that the current Israeli government, the Likud Party, does not recognize the Palestinian ties to this land or its natural resources. In fact, Benjamin Netanyahu, the current Prime Minister of Israel and Chairman of the Likud party, won the last Israeli election on the premise that the two-state solution is all but dead and that his administration will work to block any attempts to establish a Palestinian state. In many occasions he referred to the Occupied Palestinian Territories as part of Israel that cannot be compromised for any reason, much less for a Palestinian state that he vowed to scuttle as part of his election campaign platform.

This intimate link between water and national security and the widespread perception of water as a vital element in the establishment and survival of the Jewish state drives and explains Israel’s tendency to acquire more water outside of its boundaries. This relentless quest for more water even went as far north as the Syrian and Lebanese water resources and as far south as the Egyptian Nile River. To that end, Moshe Sharett, Israel’s first Foreign Minister, sums up the value of water to Israel by stating that “water for Israel is not a luxury; it is not a desirable and helpful addition to our natural resources. *Water to us is life itself*” (Rouyer 2000). Understanding the link between water and national security helps in understanding the rationale behind Israel’s negotiation strategy. To Israel, fostering its access to water resources inevitably requires maintaining control of the Occupied Palestinian Territories and the Golan Heights of Syria, while continuing its tight restriction on the Palestinian and Syrian water use (Lonergan and Brooks 1993). Based on this territorial agenda, Israel seemed hesitant to negotiate and was reluctant to reach agreements over water resources (Abukhater 2010). This is why Israel was not willing to tackle the issue of water allocation with the Palestinian negotiators and instead deferred it to the final rounds of negotiations. Any agreement or UN resolution that calls for Israel to give up part of its water supplies, even though they are acquired from other territories, or put a stop to the expansion of its illegal settlements in the Occupied Palestinian Territories and the Golan Heights of Syria are often sulfurously rejected by Israel. This plausible deniability is clearly delineated in the Likud Party Platform that holds that "immigration will be increased, and settlement will be strengthened. The decision to freeze settlements will be rescinded" (Virtual Jewish Library 1996). Until a time comes when serious changes to the current Israeli foreign policies are made or a less radical Israeli government is in power willing to take a drastically different stance in regards to land, settlements, and water resources, successful negotiations over these critical issues will continue to be in jeopardy and the prospect of a two-state solution will be surely dead as a result of Israel's plausible deniability outlined above.

2.3 The Psychological Environment and Ideological Dimension

Water issues are nuanced in nature including a multitude of issues and factors related to processes, people, and institutions that continuously and unpredictably interact at various levels and scales (e.g., local, national, regional, and even global) (Islam and Susskind 2013). In the Middle East, inter-state and intra-state conflicts related to issues of high politics dominate the geopolitical landscape and are influenced by motives and perceptions rather than actual realities. However, in this conflict there have been a number of regional and national policies often suggested by policy makers as appropriate solutions, all of which fail in one way or another to acknowledge either the multidimensional and complex nature of the shared water resources, or the contextual and perceptual factors fueling the overall conflict (Abukhater 2013). When these policies and solutions proposed on a national or regional scale, they often represent a lack of understanding of the local consumption trends that are strongly related to people's perceptions, and therefore fail to provide acceptable and implementable solutions.

In the context of the Palestinian-Israeli conflict, perceptions and reality are tightly intertwined and symbiotically interwoven into the political arena. Although for the most part an ideologically-driven conflict, the Palestinian-Israeli conflict has crucial elements that are primarily related to the psychological and ideological dimensions whose impact on the overall diplomatic process is undeniable, yet often overlooked or underestimated. In the Middle East, news, good or bad, travels at the speed of light, and the lines between reality and fiction often blur due to one's perception and ideological background. During times of conflict, these perceptions are also mixed with and amplified by emotions that can color reality with different meanings and interpretations, some of which can be irrational or far from the actual reality or the factual basis events. These perceptions, as well as the debased political discourses they are associated with, can be greatly influenced by the language employed to manufacture disingenuous images of political situations or diplomatic endeavors and their outputs (Khalidi 2013). As such, changes in general attitudes about the conflict or perceptions about each other tend to be slow and superficial (Christison 1999). Psychological and perceptual factors are strongly at play in both the current allocation as well as future plans and management of water resources in the Middle East. As such, failing to recognize the importance of perception and opting to represent water conflicts as merely a byproduct of water shortage is an oversimplification at best.

Various perceptions of land, natural resources, and of the "self" and of the "other" all influence riparian states' behavior as well as tendencies to either cooperate or altercate. These perceptions often translate into policies and actions that drive real changes on the ground. Understanding these physiological and ideological discourses is profoundly crucial in understanding the nuances of the overall conflict as well as in envisioning appropriate resolutions. Failure of policy makers to understand these perceptions on the local level and the different interpretations of what's deemed to be appropriate water management measures can inevitably produce less than holistic solutions incapable of successfully managing transboundary water resources and the perpetuation of inept policies and failed implementation mechanisms. In the Middle East, there are many conflicting psychological interpretations and ideological perceptions of the "other," the land and the environment, and the associated agricultural and landscape patterns that ensue.

2.3.1 Perception of the "Other:"

This discursive construction of the conflict, the "other" and of water resources shapes not only riparians' negotiation strategy in term of the type of issues they are willing to tackle but also their overall politics and policies in relation to water acquisition, consumption and management (Dolatyar and Gray 2000). It is an undisputed fact that the Palestinians and Israelis hold vastly different views of the "other" and of themselves, which contributed to the growing divide and the general sense of dismay and hostility. Considering the growing escalation of hostility, these diametrically antithetical views are not particularly favorable and usually involve a long list of finger pointing and accusations, and as such both parties seem to be *in pari delicto*. Both parties' view of the "other" is tailored to reflect general hostile views materializing into two forms of representation. First, communicative discourses are incorporated as part of the statecraft and national doctrine system, namely the representation of the "other" party propagated by media and educational systems. Many of the Arabs and Israelis remain captive to the idea portrayed by their media that the ongoing violence, which is represented in a cynical and cyclical way, is an inherent part of the Middle Eastern culture and an unavoidable part of life in this region (Trip'ol'II 2007). Many of the dimensions and true implications of the conflict are being censored from the media and images and narrations are being manipulated one way or another and appear to be out of sink to serve the conflicting interests of each party. With the conflict spiraling out of control, this precarious situation continues to spiral downwards as threats of a potential full-scale war becomes more and more of a possibility at any point.

Israel, early on in the conflict, has established superior discursive hegemony over the Palestinians, which allowed Israel to create an increasingly manufactured representation of reality and historic facts. This hegemonic advantage in manipulating events and outcomes reinforced and complemented Israel's military superiority, forcing the Palestinians to constantly compete in an arena whose very basic terminological rules are defined by their opponents (Khalidi 2013). To that end, Edward Said astutely explains this continued Palestinian hegemonic disadvantage by pointing out that "it is by no means an exaggeration to say that the establishment of Israel as a state in 1948 occurred partly because the Zionists acquired control of most of the territory of Palestine, and partly because they had already won the political battle for Palestine in the international world in which ideas, representation, rhetoric and images were at issue" (Said and Hitchens 2001).

Israel, through its media, often represents the whole Palestinian-Israeli conflict in a cynical way that attempts to mask (or avoid) the war crimes it committed in 1948 and 1967 and simply refers to these events as "wars." The dispossession and dispersal of the Palestinians in 1948, which they often call the "*Nakba*" (or catastrophe), has been distorted by Israel's media which represented the term strictly in reference to Israel's creation, rather than to the Palestinians' dispossession (Christison 1999). Another central element in the Israeli representation of the 1948 events is to either portray the land of Palestine as "empty" and without people, or to try to demonstrate the weak attachment of the Palestinians to their land by portraying their flight as "voluntary" rather than forced. Implying that the land was "empty" presumes the nonexistence of the entire Palestinian people and therefore attempts to exonerate Israel from any responsibly towards them. Pari passu with this "empty" land portrayal of Palestine, this "voluntary flight" has been misrepresented and rationalized by Israel's media under the pretext that the Palestinians were making way for the Arab militaries to enter Palestine and "drive the Jews into the sea." With this misrepresentation of the 1948 events, Israel, which is often shown as bearing no responsibilities for the Palestinian tragedy born out of these events, attempts to demonize the Palestinians who are frequently shown as a hostile population more bent on Israel's destruction than returning back to their ancestral homeland. This tenacious perception has been ingrained in its media representation of the conflict, as well as its foreign policy.

Along the same vein, Israel's prominent perception holds that the Palestinians have neither a rational basis for their hostility towards Israel, nor a legitimate claim to the land of Palestine. Israel also represents itself as an established sovereign state surrounded by a sea of hostile Arab nations, whose continued refusal to accept Israel's existence constitutes the underlying cause of the conflict, instead of acknowledging that their grievance towards Israel emanates primarily from the fact that Israel was established as a result of Palestinian communities' displacement in 1948. Israel in this context represents itself and its army in a purely positive light with high levels of restrain and morality, completely neglecting and contradicting the gross international and human rights violations it frequently commits in the occupied territories. This policy to conceal what Israel has been conspiring and implementing in terms of its occupation and systematic oppression of the Palestinian population has proven to be effective in deceiving and swaying its own general public and persuading it towards one direction. Keeping its population in the dark about what's happening in the occupied territories and the rule that Israel plays in keeping the status quo alive served Israel well in the past.

This complete exclusion of most of the Israeli policies and illegal activities in the Palestinian territories by the media, namely assassinations and house demolition in the Gaza Strip and settlement building in the West Bank, tends to distort and deceive the public opinion and, in turn, shape their perception about the reality of the conflict. Due in part to this inherently flawed nature of these communicative discourses and their inability to provide objective information free of ideological and political bias, the general public appear to be unable to ascertain an accurate and truthful depiction of the reality on the ground. The multitude of manipulation of facts that plague these discursive systems makes it easy for the state to skew and even shape people's perceptions about the root-causes of the conflict as well as relevant solutions to end it.

In contradistinction, the Palestinian media represents the conflict as a byproduct of a grinding military occupation from which everything else, including the construction of the separation wall and land confiscation, emanates. The root-cause of the ongoing conflict is the occupation and the failure of previous negotiation attempts is blamed on Israel's unwillingness to recognize the Palestinian right to self-determination and an independent state of their own on the 1967 land, the continued refusal to dismantle its illegal settlements and abandon its future plans to expand into further territories. They view resistance as a legitimate response against the structure of occupation of the shrinking and noncontiguous Palestinian land and the ongoing systematic oppression of its population. These communicative discourses are believed to target and manipulate public perception of the self and other's position in an attempt to paint a picture that portrays what politicians want people to think about the overall conflict, its causes, and therefore the potential for specific resolutions to serve state interests.

Perception of the "other" is also shaped by the educational system which provides another prime example of these communicative discourses in action. Palestinian and Israeli children alike have usually been taught in school two very different points of view about the history of the conflict, often mixing theology with historic events, and the other party involved in it. Employing these prominent forms of communicative discourse, including the so called "filter bubbles" and networks of mutually-reinforcing websites, social media, and news, Israel often tends to deny the term "occupation" altogether. In that sense, this denial of the term "occupation" allowed Israel to consistently refuse to acknowledge the natural, historic, legal, or religious right of the Palestinians to any part of their land by representing the occupied territories as simply part of Israel. Any narrative that acknowledges Israel's occupation of Palestinian or Syrian territories are all pained by the mainstream Israeli media with a single brush as "anti-sematic." The aim of this unrelenting campaign unleashed by the mainstream Israeli media (and eventually infected the US media as well) is to demonize and dehumanize Palestinians by delegitimizing their cause and presenting any resistance to the occupation as terrorism and that any Israeli aggression towards them is merely a response to their violence. The roots of these current ideologies adopted by Israel today can be found in the original Zionist exclusive ideologies of the of the Jewish homeland which after all made its way into the crux of the current Israeli policies and practices inside Israel and in the occupied territories. This ideology has been clearly delineated in the Likud Party Platform which asserts that "the right of the Jewish people to the land of Israel is eternal and indisputable and is linked to security and peace. Therefore, Judea and Samaria will not be handed over to any foreign administration. Between the sea and the Jordan River there will be only Israeli sovereignty. Relinquishing parts of the Western Land of Israel undermines our right to the country, jeopardizes the security of the Jewish population, endangers the security of the State of Israel and frustrates any prospect of peace" (Wagner and Walter 2014).

What's deeply troubling about this ideology that is being implemented today is that it does not acknowledge the right of the Palestinians to any part of "Judea and Samaria" and considers the establishment of a Palestinian state on the occupied territories a threat to Israel's security and future peace. As such, Israel today diametrically denies the Palestinian reaction and any resulting course of action whether it be in the form of military resistance or diplomatic effort, which is made possible by the denial of the term "occupation." Their continued denial of the term "occupation" also gave Israel a platform to expand its settlements and annex the occupied territories which are the remainder of historic Palestine comprising only 22 present. This heavily manipulated and misrepresented side of the conflict is considered one of the most nefarious aspect contributing to the prolonging of the conflict, which constitutes a major obstacle to real peace in the region.

Israel also portrays its actions in the occupied territories as essential measures to safeguard its national security, including safeguarding their illegal settlements and the increasing acquisition and transfer of Palestinian water for Israel's own use. In addition to the illegal inter-basin water transfer, building Israeli settlements on Palestinian land means that this additionally imported population needs additional water supply that is not freely available. Israel's water needs are also distorted and represented as more pressing, while Palestinian land rights and water needs are not seen as a priority to the Israeli policy makers who further developed a pejorative attitude towards them. Within the context of manipulation and influence, akin to what can be described as government-lead propaganda, much of this Israeli communicative discourse, which is distorted and entrenched in a well-orchestrated structure of misinformation and misrepresentation, may appear disingenuous when considering the historical facts and manifested reality on the ground.

Palestinians, on the other hand, view Israelis as nefarious invaders and occupiers, fought by a heroic Palestinian resistance, whose sons and daughters are more than willing to give their own lives for the cause. The story that the Palestinian people reiterated to their children is a story of displacement, dispossession and promise—a promise to survive and return back to their ancestral homeland they were driven out of in 1948 and again in 1967. This torch has been passed on from one generation to the other with an eye on achieving freedom and liberation from occupation and hope for a comprehensive and just peace that allows the Palestinian refugees to have the option of returning or just compensation. For the Palestinians, the conflict is quite straightforward and simple—ending the illegal Israeli settlement project and overall occupation are legal obligations for Israel in order to achieve future peace and for them to attain statehood and self-determination on their independent land. In response to Israel's insensitivity to their needs, Palestinians tend to develop a feeling of resentment towards Israel and its policies to prevent them from accessing their water. This resentment that continues to fester with the lack of any substantive resolution as a result of over twenty years of negotiations, caused them to lose trust in the process and each other and in turn continues to prevent them from achieving mutual understanding and cooperation. Evidently, the long standing animosity and hostility between them created adversarial perceptions towards one another which in turn lead to a great deal of skepticism of the negotiation effort itself as well as its outcome, and can potentially undermine its implementation in the future (Abukhater 2013). With this mutual lack of trust in each other as legitimate partners in peace, fueled by escalating resentment and hostility, prospects of a real peace appears to be slim to none.

Second, the historical and political interpretation of Israel's identity as a Jewish state strongly impacts the construction of water scarcity narratives and, through a multitude of institutional and techno-political arrangements, tends to provide justification for Israel's control over water resources (Alatout 2000, 2003). In this regard, the argument of water needs rather than rights or entitlement has been strongly advocated for by many actors especially Israel, which through its political, intellectual, and popular discourses, casts the argument of water rights in a negative light. This argument is utilized not only to justify Israel's exclusive control and consumption of water, but also to provide support for the many ad hoc water resolution proposals put forth for the water shortages that the Palestinian communities frequently experience. With this argument, Israel is trying to avoid the slippery slope of the Palestinian land rights which is also linked to water rights; a discussion that Israel is not willing to entertain. Although the Palestinians do not reject the idea of allocating water based on actual needs per se, representing the Palestinian water crisis as simply a matter of absolute need in separation from rights is an unfortunate misrepresentation of their real issue, which remains to be a major concern to them (Abukhater 2010). Acknowledging the Palestinian water rights is essential from the Palestinian perspective and should not be ignored as it personifies the link between the right to water and land, both of which have been denied by Israel. This Palestinian perception, which provides a radically contradictory ideology to that of Israel regarding their water use, asserts that "Israel is using the Palestinian water." Saeb Erekat, a high official in the PNA, reiterated the Palestinian belief that water is part of their occupied land by stating that "The Israelis are stealing our water and it must stop. They must be reminded that they are sitting on our chests; and not by an act God, but by an act of war" (Rouyer 2000). This aspect of the ideological and psychological environment must be acknowledged and reconciled when considering future water allocation schemes that are based on pure needs rather than rights. The immediate water needs for the Palestinian communities must be fulfilled but it should not be viewed as the end of their story. Palestinian communities, including refugees, are entitled to water access rights that emanate from their inalienable right to their land. Any resolution that is not aligned with this thinking will not be positively viewed and as a result will risk rejection and ultimately failure.

Regardless of their accuracy, truthfulness, or legitimacy, these perceptions of the "self" and the "other" (or "us" versus "them") invoke a state of constant fear and ungrounded distrust and resentment that can fuel hatred of the "other" and therefore sustain the conflict. Because of these perceptions of one another propagated by this evolving information environment where powerful actors work to promote lies and discredit the truth, the Israelis and Palestinians do not see the other party as a partner in peace but rather an adversary. This creates a vicious cycle which continues to widen the communication and trust gap and leads to skepticism of, and hesitation to entertain, any attempts for reconciliation as long as they are unilaterally proposed and initiated by the "other" adversarial party. This calls for the intervention of a mutually trusted third-party to act as a mediator to promote cooperation and help the parties to broker a peace settlement ending decades of hostility, distrust and conflict, as we will discuss in the last chapter.

2.3.2 Different Distortive Representations and Interpretations of Terminologies

This conflict provides a classic example of two parties holding dramatically diverse and conflicting ideological stances, where vastly different interpretations of key concepts and representations can be found. The role of language usage and the media to describe and interpret an action or a situation is extremely prevalence in this conflict. Specific distortive terminological rules that define the official lexicon in this conflict can sway the populous opinion to serve disingenuous political agendas. This slovenliness of language to serve political propaganda makes it easy for state actors to misrepresent facts, manipulate the truth, and spread corrupted phraseology to profoundly obscure reality (Khalidi 2013). As George Orwell aptly put it: "if thought corrupts language, language can also corrupt thought. A bad usage can spread by tradition and imitation even among people who should and do know better" (Orwell 1946). For example, terms such as "peace process," "self-determination," "security," "terrorism" are heavily loaded and are used to describe certain actions or groups in order to corrupt thought and cloak reality (Khalidi 2013). The use of the term "terrorism" has been employed in the Israeli media to *exclusively* reference the actions of the Palestinian military groups, and cannot be used to describe any of Israel's military actions which resulted in the killing of thousands of Palestinian civilians. The term "security" has also been manipulated in this lexicon to reference Israel's existential fears and therefore is blindly regarded as an absolute priority that takes precedence over anything else regardless of importance to the peace process or Israel's actual security (Khalidi 2013).

Similarly, the concept of "efficiency" in relation to water usage, for example, has been subject to different contradictory interpretations depending on who defines it and their ulterior motive. It is no surprise that the Palestinians and Israelis have different representations and interpretations of the term. "Efficiency" for Israel, on one hand, means "more crop per drop" regardless of how much water is used. This interpretation of the concept of "efficiency" has its roots in the notions of "productive efficiency" and "allocative efficiency" (Frisch 2000). "Productive efficiency" is linked to, and strongly influenced by the technological advancement of Israel's water management practices that aim to obtain a greater yield from a unit of water, whereas "allocative efficiency" emerges from the market-driven water allocation and pricing system. This understanding of "efficiency" as "productive efficiency" promotes the search for sound water management to increase supply or decrease demand. Efficient agricultural practices that Israel utilizes include drip irrigation, water reuse (which ostensibly constitutes 70% of Israel's municipal water supply), desalination, greenhouse technology, cloud seeding, and genetic bioengineering of crops for drought and salinity resistance (Wolf 1996). "Allocative efficiency," on the other hand, is linked to boosting economic values and production being generated. For Israelis, their water consumption, although may be considered excessive by many standards, is viewed as efficient based on the aforementioned criteria of economic-based efficiency. As such, Israel is often regarded by many western countries that share similar economic-based criteria for efficiency such as the US, as a leading example of sound water management. This view, which glamorizes Israel as a global leader not only in solving its own water crisis but also the global water problem, holds that "Israel could not have made the desert bloom without its incredible innovations in water technology. As the world becomes more aware of the importance of conserving water, they are turning to Israel for exports and expertise" (Hazony 2015).

These misrepresentations of Israel and its water management practices seem to be oblivious the fact that most of Israel's water supply is acquired in the Palestinian territories and further fail to distinguish between water management practices on the micro and macro-levels. Evidently, Israel's water use can be characterized as excessive rather than conservative when one considers that a significant portion of Israel's water supply comes from the occupied territories and later transferred into Israel for its use. In addition, Israel maybe successful in using its expertise and technology in managing water use efficiently on the micro-level, namely at the consumer level. However, Israel suffers severely from lack of efficiency of water allocation on the macro-level, posing a great challenge for the country and the region as a whole, which seems to be confronted with an acute state of vulnerability to water shortages now and in the near future. One of the few examples of demand-side management strategies that Israel relies on to alleviate pressure on natural resources and maximize "efficiency" is restricting water consumption for the end user. This, along with the fact that Israel's per capita water consumption constitutes 25% of that of the US, provides fuel to the claims that promote Israel's water allocation strategies as conservative and efficient (Lonergan and Brooks 1993). However, evidence shows that except for rare occasions of conservation, typical Israeli policy in water resources management relies more on increasing the supply than decreasing the demand by heavily subsidizing water. Actual figures of water consumption data indicate that Israel still consumes great volumes of water for irrigated agriculture compared to its neighboring countries located in the same region. This is partially due to its national policy in handling water distribution and pricing system, which seems to paradoxically contradict with the very aim and definition of both "productive" and "allocative efficiency." The problem here seems to be a byproduct of Israel's policy of subsidizing water for agricultural irrigation (Lonergan and Brooks 1993). This supply-side policy that promotes more water use not less cannot be deemed efficient, if efficiency means conservation of resources. This policy tends to increase water consumption in the long run and undermine both "productive" and "allocative efficiency" by counteracting conservation and market economy goals. Nevertheless, the key issue is not whether Israel is reducing its water use per se, but rather whether this reduction in its water usage is sufficient enough to match the magnitude of the crisis facing its growing population and the region at large. Evidence points to the contrary.

The meaning of "efficiency" for the Palestinians, on the other hand, is grounded in their cultural practice and religious belief and directly translates into conserving water at all times. Their general water allocation policies tend to lean more towards leveraging the overall guiding principles of demand-side management (DSM) and water conservation. Realizing the importance of water conservation as a basic religious principle, the Islamic teachings encourage people to conserve water even if a running river is present (Faruqui 2000). This is because Muslims believe that the earth and its resources are not a gift that they inherited from their grandparents for their unconditional use, but rather a sacred trust that has been placed in their hands for safekeeping so it can be also passed onto the future generations. Water in particular is deemed as a blessed and sacrosanct resource sent from God to purify and cleanse the earth. Water has been mentioned in the Holy Qur'an many times in the context of reviving the earth while casting safety and peace. To that end, the Holy Qur'an (in Surat Al-Anbiyaa "The Prophets") states: "Have those who disbelieved not considered that the heavens and the earth were a joined entity, and We separated them and made from water every living thing" (Qur'an, 21:30). The Holy Qur'an (in Surat Qaf) further explains: "We sent down water from the sky, blessed water whereby We caused to grow gardens, grains for harvest, tall palm-trees with their spathes, piled one above the other—sustenance for (Our) servants. Therewith We gave (new) life to a dead land. So will be the emergence (from the tombs)" (Qur'an, 50:9–11). Following these teachings, the Palestinians regard water as part of their land, identity, and religion. As such, water is not a commodity to be expropriated, but rather a precious resource worth saving in their view. These teachings also promote the belief that humans are equal not only in terms of rights, but also responsibilities and obligations towards one another and the environment. Basic human rights include securing equal access to a clean and reliable source of water to all people, rich or poor, and regardless of their status. This concept of basic rights also encapsulates a prioritization mechanism that takes into account the right of all living things including animals and plants, which are entitled to a fair share of water rights.

In addition to rights, there are a number of unspoken moral and cultural principles that are clearly understood by everyone to guide the overall perception of what is considered efficient water consumption and management practices. These include moral and ethical responsibilities related to demand-side management by maintaining the integrity of the shared water resources and preventing waste. For example, only two to three liters of water is the recommended amount for a sufficient bath by the Islamic and societal standards (Faruqui 2000). People are also encouraged to share water and prohibited from withholding it from others, especially travels, who need it the most. The process of water allocation is, therefore, solely and exclusively based on respecting and catering to all parties' needs, including the right of the environment. Balancing human and environmental water needs is also crucial in maintaining human health and dignity as well as a healthy ecosystem. In pursuit of this goal, Sandra Postel, 1997, in her book, Last Oasis, astutely writes: "We have been quick to assume rights to use water but slow to recognize obligations to preserve and protect it…. In short, we need a water ethic—a guide to right conduct in the face of complex decisions about natural systems we do not and cannot fully understand." This unique cultural practice guided by a set of environmentally-friendly interpretations of the concept of "efficiency" is closely linked to principles of water conservation and entails proportional distribution of benefits as well as costs of water use among all parties.

2.3.3 Perception of the Land and the Environment

Israel's agricultural policy and water management practices have been heavily influenced by the Zionist ideology whose main goal is to create and sustain a Jewish homeland within the geographical context of Israel/Palestine. Given the heavy ideological and perceptual aspect of this conflict, which appears to be more prominent than in any other disputes, Zionists view Israel as a "safe haven" land that has been promised to the Jewish people by God. The acceleration of their immigration to Palestine is also put in the context of "the right of return" to their "promised land." This theologically-driven perception of Palestine as the "promised land" shaped their tenacious view of the land and its natural resources as being sacred and nonnegotiable. Water is a central component of the Zionist ideologies and ethos of nation building (Rouyer 2000). For Zionists, the "ideology of agriculture" is at the core of their belief system and a crucial part of their core identity. Israel often promotes its image as a modern nation whose pioneers were able to make the desert bloom by utilizing sound water and irrigation management practices (Christison 1999). To make this vision of "Eretz Yisrael" (a sovereign Jewish state) a reality, large-scale agricultural development is regarded as an essential component to accommodate the large numbers of Jewish immigration from around the world. Because water is the lifeline of irrigation and agricultural production, Israelis view it as a cornerstone for the establishment, survival, and security of their sovereign Jewish state. The perceived value of water in this context extends beyond its basic economic value to including a much more significant ideological and political importance to Israel. This is an endlessly expansive notion that includes a multitude of issues and factors that shape the current conflict and potential future settlement. The Zionist common use of the corrupted phraseology "making the dessert bloom," which insidiously implies that before the arrival of Zionists, Palestine was a wasteland, provides Israel with a discursive hegemony that aims to deny the existence of the advanced Palestinian farming communities that inhabited the land throughout the past centuries (Khalidi 2013). This has a profound impact on obscuring the reality and a far reaching implication to indemnify their assertion and quest for "a land without a people for a people without a land."

The "ideology of agriculture" and the need to keep the land "green" is what motivated Israel to expand its illegal settlements into the occupied West Bank and East Jerusalem by establishing a rural form of human settlement that prominently depends on farming practices, called the *Kibbutz*. This expansive presence is perceived as a national goal of high importance for Israel and its survival. This policy, which favored the expansion of agricultural irrigation and water development projects, also helped in strengthening the perceived ties of the settler community to what they consider to be their Jewish homeland. In this ideological context, water is regarded as an essential element not only for accommodating the growing Jewish population within the State of Israel, but also for fostering their ties to "Eretz Yisrael." This ideological perspective made it easy for the State of Israel, through a centralized system of water acquisition, management and allocation, to maintain exclusive ownership of all water resources within the country, including the occupied territories. This prominent theological ideology continues to drive much of Israel's decisions in the occupied territories as well as in relation to its foreign policy.

This ideologically-driven push to keep this semi-arid land green means that Israel needs to acquire as much water as necessary to farm and irrigate it, which motivated Israel, in the 1950s, to establish its 130 km-long National Water Carrier, referred to as the Kinneret-Negev Conduit, that transfers water from Lake Tiberias and delivers it to the rest of the country penetrating into the depth of the Negev desert to support large-scale agricultural irrigation projects (Shamir and Haddadin 2003). However, in light of the fact that agriculture is considered a minor sector in Israel's economy, the reason for keeping the land green seems to have little to do with agricultural production and economic development. The establishment of this National Water Carrier, shown in Fig. 11, would not have been possible without Israel's advantageous hydro-strategic position with the annexation of the Golan Heights in 1967, which allowed it to secure access to the principal headwaters of the Jordan River. By continuing to avoid entering any peace negotiation with Syria over the Golan Heights, Israel continues to not only utilize those water resources, but also prevent Syria from diverting more water from the Yarmouk River and its main tributaries. By so doing, Israel ensures its exclusive control over Lake Tiberias by maintaining its control of the Golan Heights, which due to its strategic location provides an example of how Israel's perception of land, and by extension water resources, shapes its foreign policy decisions as well as its relations with neighboring countries.

The ideology of agriculture also provides a reasonable explanation of Israel's relatively high water consumption rates, which Israel often attempts to justify using

Fig. 11 Israel's National Water Carrier. *Source* Illustration rendered by Michael N. Erickson, 2018 based on data from the Britain Israel Communications and Research Centre, BICOM (2007)

the argument that it is a more developed society that requires higher amounts of water. In this context, Israel promotes the view that the Palestinian water scarcity problem can be solely blamed on infrastructure mishaps. Their representation of the problem is distorted and portrays the Palestinian water crisis simply as a byproduct of the lack of adequate infrastructure capable of extracting greatly increased volumes of water to meet their exponentially growing demand.

The Palestinian perception, on the other hand, is mainly shaped by their previous experiences in 1948 and again in 1967, reflecting their fear of dispossession and the 20th century Jewish immigration (Rouyer 2000). They view the origin of the conflict to be mainly related to the events of 1948 that resulted in their subsequent forced flight, known to them as the "*Nakba*" or catastrophe. Their view of their land as being occupied since 1967, namely the West Bank, the Gaza Strip, and East Jerusalem, shaped the way Palestinians perceive and interact with Israel, the land, and its natural resources. Living in a predominantly agricultural society, water is seen by the Palestinians as an extension of their land and an integral part of their daily life and struggle for independence and self-governance. Their view of the land and water is not only confined to their basic economic values, but it rather personifies a highly tight bond between them as a people and their occupied land viewed as part of their identity and sense of belonging. They believe that without equitable allocation of land, water and natural resources peace will remain elusive.

These conflicting and vastly different ideological points of view are also evident in the ensuing landscape patterns that can be seen on both sides of the boarder. Evidently, landscape patterns reflect not only states' national identity but also how people actually use and manage water for irrigation and agricultural purposes, both on the local and national levels. These perceptions and meanings of the land, the environment and natural resources were clearly documented and translated into distinguishable landscape patterns on the ground representing the people, their respective ideologies and cultural and societal practices.

By examining the agricultural and landscape patterns in both Israel and the Gaza Strip, for example, one can clearly see the evidence of the different agricultural water usage and management practices and therefore understand their geopolitical and hydropolitical implications in this region. As depicted in the satellite imagery shown in Fig. 12, the landscape pattern that is prevalent along the eastern border fence of the Gaza Strip reflects small and fragmented farmlands and a more conservative use of water. This pattern of landscape connotes a small town feeling and a gradual transformation from urban to rural areas exhibiting indigenous patterns of landscape and farming practices that dominated this land pre-1948. As one moves to the west, however, the landscape in the rest of the Gaza Strip shows a highly dense area composed of overwhelming amounts of concrete (shown clearly in the grey color of the map, which can be easily spotted and delineated) with very little green areas. There is no surprise here as the Gaza Strip is believed to be the world's most densely-populated areas with the majority of its population are refugees. The intriguing part of this picture lies within the narrow corridor along its eastern border fence with Israel which shows a stark difference between the two areas in terms of population density and landscape patterns. It reveals a troubling secret about the inequitable water access. One side, Israel, has more access to water displaying it in the form of green and lush landscape, while the other, Gaza severely lacks that access and shows it in the dryness of its land.

This indigenous agricultural and landscape pattern that dominates the Palestinian terrain represents more conservational water use reflecting the overall perception of the value of water and the need to preserve it. Bearing in mind that this is a semi-arid region, such indigenous mode of landscape still manages to thrive without harming

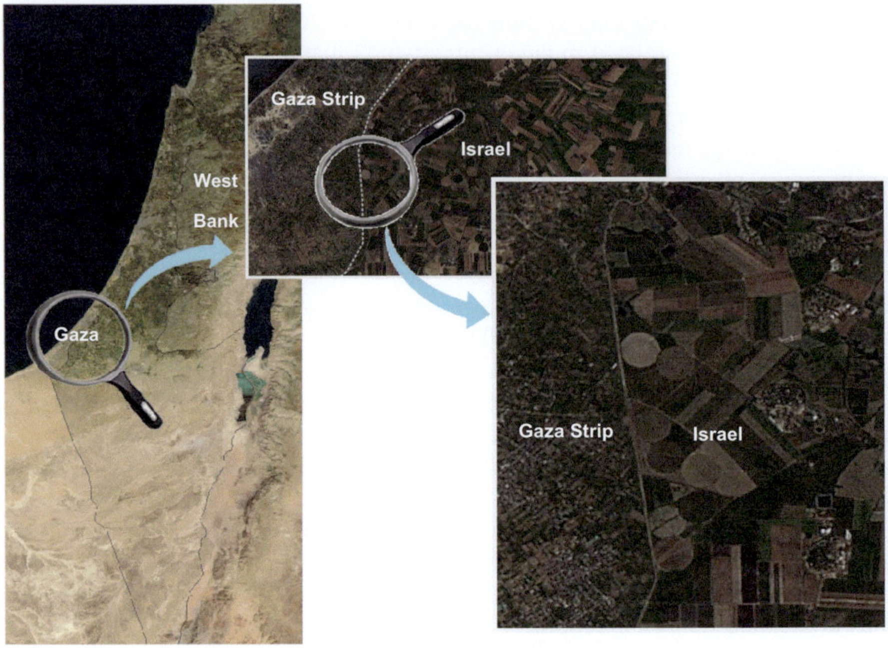

Fig. 12 Agricultural and landscape patterns in Israel and Palestine. *Source* Google Maps

the environment or the need for extensive water use that tends to deplete the limited resources. Through careful water use and irrigation management, the Palestinian landscape includes types of vegetation that are more resilient to drought conditions and can survive in this semi-arid environment. Small trees, such as olive and citrus that are interwoven into the urban fabric made of primarily small towns and villages are considered the hallmarks of the traditional landscape of Palestine, as shown in Fig. 13.

On the other side of the border fence, the picture is quite different with the introduction of an alien reality that reflects a vastly distorted depiction of the land and the environment. In addition to the deliberate destruction of over 530 Palestinian villages to conceal the reality and magnitude of what conspired in 1948, the subsequent introduction of an artificial and foreign form of landscape to mimic the lush European countryside, from where the Jewish immigration emanated, represents an attempt to eliminate and mask the traditional landscape that defines the small Palestinian villages and their cultural identity. Their goal is to overwhelm the general public with a torrent of misinformation and a myriad of manipulative discursive construction of the conflict and the land itself in a devious concoction to delegitimize the Palestinian argument of a homeland and therefore discredit their assertion of the *Nakba* events. This manufactured landscape contributed in altering the traditional character of the land, the appearance of its landscape and defining topographic elements and the way it is viewed as green and lush hills diametrically opposing the Pales-

Fig. 13 Formal structure of the Palestinian landscape—olive trees and small towns. *Source* Susiman et al. (2005)

tinian representation of their ancestral homeland and its countryside. As a result, the Israeli landscape shows *prima facie* evidence of large-scale agricultural development foreign to this climatic region. This alien landscape requires the adoption of high technological advancement in farming and irrigation along with extensive water use to sustain—water that is not always abundantly available in this semi-arid region, putting more pressure on the available, already strained and disputed water resources.

2.4 *The Political and Legal Dimension*

2.4.1 Political and Diplomatic Dimension

Needless to say, considering its hydropolitical significance, water is one of the most critical issues that shape the geopolitical landscape of the Middle East. In this semi-arid region, water is a deemed a precious resource that has serious implications to other high politics issues. As such, its role in shaping various political interactions in this region cannot be overemphasized. Given its political importance in the Palestinian-Israeli conflict and its potential resolution, water remains an unresolved challenge facing the entire region today and was one of three major issues deferred to the final rounds of negotiations. This can be attributed to a number of factors that underline the overall hydropolitical conflict. First, the current status quo of Israel's water acquisition and use indicates significant reliance on water resources originated in the Palestinian territories, namely the West Bank rain-fed Mountain Aquifer, which is the main source of domestic water use for Israel. This hydrodependency of Israel on Palestinian water resources makes Israel less than willing to entertain any negotiation with the Palestinians over these water resources that it currently utilizes. Any meaningful resolution will inevitably result into a transfer of these resources to the Palestinians, leaving Israel short of at least one third of its annual water supply which is already insufficient in the face of its increasing water demand exacerbated by its growing population (Lonergan and Brooks 1993).

Second, the fact that both populations depend primarily on the same over-extracted water resources for their very survival created a situation known as zero-sum game, where one riparian state's gain is perceived as another's loss. In this situation characterized by unprecedented degree of regional water stress and mistrust, Israel appears as the winner in this game by benefiting from keeping these resources under its full control. This rationale that Israeli negotiators adopt implies that it is in Israel's best interest to prolong the status quo and avoid the inevitable water negotiations with their Palestinian counterparts. This particular aspect of securing uninterrupted water access for Israel greatly contributed to the difficulty in initiating and reaching agreements over water. Ironically, this approach of continuing to monopolize the Palestinian water resources, however, may evidently be more costly for Israel when compared to other alternatives, and considering the ongoing confrontation with the Palestinians that these Israeli policies trigger, which often requires the deployment of its additional military and intelligence resources. Many scholars in this field argue that it would be a more economically feasible strategy for Israel to engage the Palestinians in a bilateral negotiation process over water and reach a plausible agreement. However, Israel, unwilling to immediately tackle the impending water issue until its security is assured, seems to opt for the second most costly option of confronting potential conflicts with the Palestinians. As we discussed earlier in this chapter, one of the aspects of this conflict that can amply explain this anomaly in Israel's policies in water allocation is related to the fact that the importance of water for Israel is not simply a matter of economic significance or agricultural production. Rather, Israel's water policy is solely driven by political, religious, and ideological motives.

Third, the currently available water resources is scarce and inadequate to fulfill the current needs of both parties to begin with. As such, both parties will continue to experience severe water shortages unless they collaborate to develop new large-scale alternative approaches that would either augment their respective access to new water resources or conserve their current use of available resources. This alternative could be introduced in the form of large-scale demand side management (DSM), by conserving water, or in the form of supply side management (SSM), by finding new sources of water whether it be technologically-driven sources, such as desalination of seawater or brackish groundwater, wastewater reclamation, or transboundary water imports. Either way, these approaches must be large and significant enough to provide a solution proportional to the magnitude of their current water crisis. Fourth, a major deterrent to negotiation is partly related to the fact that the practice of modern negotiation and conflict resolution is relatively new to the Middle East, and ipso facto are not favorably viewed by many (Irani 2000). This is one of the key hurdles that emerged from the previous attempts to settle the dispute, all of which lacked acceptability as they were seen as categorically disconnected from the local Middle Eastern culture and its indigenous practices. Akin to approaches that demonstrate a gap between public needs and public policy, these previous negotiation approaches were perceived as a western panacea, which prevented the parties from advancing their negotiations towards more plausible agreements and robust implementation on the ground.

2.4.2 Inadequacy of International Law

Although the impact of international water law on this conflict cannot be overstated, the law, which is interpreted differently by the opposing parties, is surprisingly not fully implemented in many cases. Often times both parties, namely the Palestinians, ground their arguments in relationship to their respective claims to the land and water on principles of international law. These laws address the issue purely as a matter of sovereignty over water resources (i.e. who has the right to what resources). These key principles applicable to the Palestinian-Israeli water dispute are for the most part outlined in the Hague Regulations of 1907, the Fourth Geneva Convention of 1949, the Helsinki Rules of 1966, and the 1997 UN Convention on the Non-navigational Uses of Watercourses.

While it introduces some elements of water needs, the international law is more concerned with issues of sovereignty over water resources rather than actual needs. These laws often attempt to provide general guidelines and elements for distributive justice in relation to water resources allocation (Abukhater 2013). In that sense, access to water resources is prioritized and protected based on uses and users. In handling the conventional long-standing dispute between upstream and downstream riparians and existing and future uses, the international law provides more protection for the downstream users and existing uses (Wolf 1999). In practice, however, applying these principles is often conducted with little regard for environmental concerns. Although it laid some ground rules to manage the sharing of the disputed water resources, international law seems ambiguous and in many cases contradictory. In addition, the law does not provide a clear enforcement mechanism to ensure accountability in applying and implementing its rules.

On the whole, common water allocation practice is based on either hydrographical or chronological grounds. The hydrographical argument is grounded in two irreconcilable arguments including "the doctrine of absolute territorial sovereignty" and the principle of "territorial integrity," both of which are concerned with the origin and spatial configuration of a river or aquifer in relationship to each riparian state's boundaries (i.e. where the resource originates and how much of its boundaries resides in each riparian state). Upstream riparians, such as Israel, often rely on the Harmon Doctrine, reflecting the absolute territorial sovereignty argument, to justify their absolute sovereignty over the water falling in their boundary (Spiegel 2005). Based on this claim of absolute sovereignty, Israel diverts over 75% of the Jordan River's water before it reaches the Palestinian West Bank which is a downstream riparian, who based on the principle of "territorial integrity," usually argues that Israel should not alter the flow of the river. On the other hand, the chronological argument is based on "prior appropriation" and reflective of the historical use of water (i.e. first in time, first in right) (Wolf 1996). The Palestinians, as a downstream riparian, also reference this historic right argument to support their access to the Jordan River water and further criticize the excessive Israeli use that impedes the flow of the river and their ability to meet their basic water needs. These aforementioned legal principles, which are concerned with issues pertinent to sovereignty, absolute river integrity, and historic rights are considered "extreme" because of their strict regulations and lack of flexibility or consideration for principles of equitable allocation (Wolf 1999; Giordano and Wolf 2001).

Regardless of what international law is applicable to this conflict, the issue of far reaching interest remains related to the question of water allocation equity. These laws introduced various measures to promote water-sharing equity, including rights-based measures, needs-based measures, and measures based on economic grounds or efficiency. However, the applicability, adequacy, and effectiveness of these conflicting doctrines pertaining to the Middle East water resources is heavily disputed and often in question by one party or another. The question of equitable water allocation has been moderately addressed by the international law, though not adequately delineated or fully implemented. Laws such as "the doctrine of limited territorial sovereignty," "the Helsinki Rules of 1966," and "the 1997 UN Convention on the Non-navigational Uses of Watercourses" are referred to by Wolf, 1999, as "moderate principles" due in part to the fact that they focus on issues of equitable use and the obligation not to cause significant harm.

Pertaining to the question of rights-based equity, international law gives priority to both "existing" and "historical" uses of water (Kally 1993). However, trends of past agreements show a pattern of underscoring the significance of historical and geographical considerations in international water disputes (Wolf 1999). The Palestinians, based on their historical uses of water, assert that their rights to water were always protected during the Ottoman, British, Egyptian and Jordanian administrations. It is only under the Israeli occupation have these historical rights been summarily denied (Rouyer 2000). The heart and soul of the Palestinian legal claim to water resources of the Gaza Strip and the West Bank rests on the international law of "belligerent occupation." Consequently, the Palestinians criticize Israel's practice of exploiting surface and groundwater in the occupied territories as being illegal and a gross violation of international law. Conversely, attempting to indemnify its actions, the Israeli government often denies the applicability of these laws by denying the term "occupation."

The Helsinki Rules of 1966, which put emphasis on the right to "beneficial use" rather than the right to water per se, provided eleven factors pertinent to hydrographic and socio-political criteria. Those factors are recommended as a useful legal framework to resolve water allocation between the Palestinians and Israelis, as will be discussed later in this book.

The 1997 UN Convention on the Non-navigational Uses of Watercourses provided general guidelines that both the Arabs and Israelis could greatly benefit from, when meaningfully considered. The two main principles introduced in the 1997 UN Convention are "equitable and reasonable utilization," set forth in Article 5, and "no significant harm" to other watercourse states, set forth in Article 7 (Diabes-Murad 2004). It further gives "special regards" to the "requirement of vital human needs" (Wolf 1999). Although the concept and guidelines for "reasonable and equitable" sharing of common waterways were introduced in the Helsinki Rules of 1966, as well as the 1997 UN Convention, no clear definition of this concept was ever provided, making it subject to various interpretations (Caponera 1985). This ambiguity associated with certain key concepts makes it easy for disputing parties to interpret the law differently according to their own perceptions and ulterior motives. It is imperative to point out that international law provided general rules to allow room for flexibility, adaptability and discretion for states to consider and evaluate alternative solutions for various cases. However, in the case of the Arab-Israeli conflict, this flexibility caused more disagreements among disputing parties over the meaning and interpretation of the law to justify and legitimize their positions. In the case of the Palestinian-Israeli water conflict, this aspect was particularly problematic and continuous to cause a great deal of discrepancy regarding the basic ground rules for the negotiations.

International law is an area of great controversy that is consistently hewn by Israel's denial of its applicability and therefore its disregard of its principles on the basis of irrelevance or intrusiveness. Although binding on all nations, the applicability of many elements of these laws to the Arab-Israeli dispute is denied by Israel based on its denial of the term "occupation" as discussed earlier (Rouyer 2000). Israel, not fulfilling its obligations as a "belligerent occupier," has been unilaterally in control of all water resources of the Occupied Palestinian Territories as well as the

Syrian Golan Heights. Therefore, the current utilization of these international water resources does not comply with the international law, namely the "equitable and reasonable utilization" and the "no harm" principles. This is especially true considering the significant hardship this Israeli control and utilization of Palestinian water imposes on the Palestinian communities, resulting in dramatically different water use patterns as demonstrated earlier.

The applicability of the international law to the volatile watercourses of the Middle East is inconclusive. Enforcing the law however is crucial in giving it a more impactful and puissant role in resolving this conflict. Without a clear enforcement mechanism of these applicable laws, evidence suggests that the prospects of peace will be slim. Aside from these laws, the focus should be on finding common grounds for equitable distribution of water resources based on the real needs of the population, which should stem from water rights, rather than the desires of the riparian states. Regardless of water ownership and entitlement, satisfying the urgent need for domestic and potable water should have priority over agricultural application. In short, the international law is necessary, yet not sufficient in and of itself. Goodwill on both sides, the intervention of a third party, and an ongoing negotiation and cooperation process are warranted for the success of effective and environmentally-sound water allocation practices.

In conclusion, this chapter provided a framework for analyzing water related disputes based on a generic formula that can be adopted in various cases. This formula was also used to analyze the Palestinian-Israeli conflict and provide a diagnostic account of the overall geopolitical landscape including problems facing the region, in general, and more specifically suggest recommendations towards potential resolutions. The overall geopolitical atmosphere can only be characterized as volatile, constantly on the verge of outright war. The fact that the water issue remains unresolved impacts the political arena and exacerbates the conflict. As such, water plays a crucial role in sustaining the conflict and any future resolution must be cognizant of the importance of resolving these water issues first. The deferment of water negotiation means that not only will the water scarcity crisis remain unresolved for both Israel and Palestine, but also the current status quo will continue, creating a situation of severe inequity that will continue to disrupt and hinder any chances for peace. With the continued refusal to proceed with negotiations and the ensuing limited channels of communication and diplomatic effort, the relationship between the two parties can be characterized as unidirectional, where Israel, looking out for its own interest as the most powerful state, causes environmental harm to the Palestinians, but manages to inflict no or little harm to itself in the process.

The potential to utilize impactful hydro-diplomacy via equitable water allocation as an entry point and prerequisite to resolving "high politics" issues, such as land and sovereignty, is crucial in amicably resolving this conflict. The process as well as outcome of any future negotiations must recognize the inherent complexities and nuances of the Middle East geopolitical reality that cuts across various environmental, hydro-hegemonic, psychological and legal dimensions. If all sides in this conflict mange to work together to discover and secure new sources of water to equitably and adequately meet their respective demands, sustainable peace can be attainable. This is because water, a simple substance, is very crucial to the survival of both nations. This is especially the case since the larger political and cultural clash is intertwined with the problem of water shortage, all of which makes reaching a meaningful and practical resolution almost insurmountable. Without equitably resolving the water crisis of the Palestinians, Israel's own interests, security, and stability will continue to be in focus. Attaining and maintaining water equity and security for both parties must be utilized as a catalyst for attaining and maintaining the overall regional stability. For any negotiation attempt to succeed in producing impactful, scalable and acceptable resolutions, it must first recognize and leverage the link between the two aspects, on one hand, and the importance of fostering equitable water sharing, on the other. Otherwise, the perpetuation of the current peace process which has proven to be more focused on the process itself rather than a real substantive outcome would most likely produce more delinquency in finding a final resolution and the perpetual fanning of an already volatile situation.

Chapter 3
A House Divided—The Requiem for the Two-State Solution

United we stand, divided we fall.

—AESOP

This chapter casts light on a critical issue that continues to be a major obstacle to achieving a united vision for peace which is related to the internal Palestinian divide, namely between the Gaza Strip and the West Bank governments as well as the perception of Palestinian populations living in these two regions of Israel and of each other. This divide, which has been widely neglected or overlooked by many scholars, has multiple dimensions that are traced back to external intervention from Israel to widen and exploit this demographic, cultural and political division and the obvious noncontiguous nature of these two geographic locations. There are also important aspects related to the origins of these two populations, differential political affinities, and other environmental aspects and cultural practices that widened the aforementioned divide. This is a critical chapter in the overall picture that the book tries to present because it challenges the validity and practicality of the two-state solution and further paves the way for the fifth chapter that would consider these aspects of the Palestinian house as a core component to be addressed in the final resolution.

Palestine; one land, one country, one people for thousands of years. Today, however, the situation is quite different. Palestine has been always one country with Palestinians living in it in harmony for many generations. As was discussed in the first chapter that traced the history of the conflict and the gradual cartographic regression of Palestine overtime as a result of Israel's expansive policies, it was not until the middle of the twentieth century that Palestine was severed into two smaller geographic areas that are completely separated and isolated from one another—the Gaza Strip and the West Bank.

Home to almost two million people, the Gaza Strip (or simply Gaza) is a 360 km^2 strip of land on the eastern side of the Mediterranean Sea bordering Egypt on the southwest and Israel on the east and north sides of the border fence. This makes it the most densely populated place on earth. The original Gaza Strip used to extend over a much larger area than its current boundaries. A significant portion of the territory is also confiscated and depopulated due to a buffer zone that has been implemented

A. Abukhater, *Palestine - Peace by Piece*,
https://doi.org/10.1007/978-3-030-03985-1_3

and maintained by Israel and is completely housed within the limited Gaza borders encroaching on the already confined area of the Strip making it even smaller for the large population currently residing in it. Home to over 1.7 million people, the West Bank is a five and a half thousand square kilometer territory west of the Jordan River. Until recently, these two geographic territories have always been one country, known as Palestine. These two areas have always been connected and constituted key parts of what has always been known as the land of Palestine, even under the Ottoman Empire and later the British Mandate on Palestine, as discussed in the first chapter. Today, these two geographic entities are not only two separate areas but are also almost completely segregated from one another. How did a united country get divided, compartmentalized and torn? What preceding events led to this division? These questions can only be answered by following the chain of events throughout recent history and understanding the context within which these events emanated. It is crucial to note, however, that the following examination of key relevant events is not intended to provide a comprehensive historical account of the region, but rather contextual understanding of key events leading to the current Palestinian divide.

1 Early Twentieth Century and Six-Day War

After World War I, Palestine remained one country but was colonized by Britain from 1922 until 1947. During these years, the Palestinian fabric stayed intact and remained stronger than ever to confront the British invaders and ostensible complicity with the Jewish newcomers (Fig. 1).

Finding itself in *flagrante delicto* with its collusion with the Zionist movement's agenda and ongoing colonial policies in Palestine, Britain was eager to completely withdraw from the then Palestine. As a result of several decades of political pressure and the ongoing and increasing violence it created throughout the country, Britain referred the problem to the United Nations. On November 29, 1947, the UN Resolution 181 was passed calling for an end to the British Mandate of Palestine no later than August, 1948, and the creation of two separate states, one for the Palestinian Arabs and the other for the Jewish people, as shown in Fig. 2.

This UN resolution partitioned Palestine, for the most part on equal basis, into seven areas—three Jewish areas and three Arab areas with Jerusalem under international control. Akin to many other nations in the Middle East, Palestinians have been seeking independence and self-determination since the beginning of the twentieth century. Rather than affirming their self-determination by utilizing a more democratic approach to resolving the issue, the newly formed United Nations instead used another approach by which an unexperienced outside agency proposed to divide the country into these two Palestinian and Jewish areas. As was discussed in the first chapter, this plan was rejected and was ipso facto never implemented. The rejection of the plan was due to its irrefutable failure to affirm the right of the Palestinians to self-determination and the fact that it effectively proposed a two-state solution on predefined areas, disregarding the ethnic and demographic composition of the

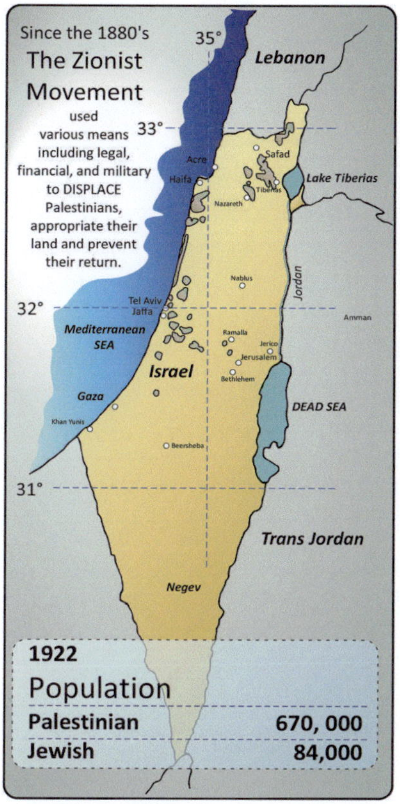

Fig. 1 The map of Palestine in 1922. *Source* Illustration rendered by Michael N. Erickson, 2018 (adapted from Visualizing Palestine)

country. It also allocated more than half of Palestine to a relatively small Jewish population when compared to the Arabs who constituted the majority of the population and owned most of the land. It was also problematic regarding the geographic separation it created for the city of Jaffa which was segregated from the rest of the proposed Arab state and the Gaza Strip which was also separated from its farmlands. The plan which was endorsed by the U.S., was accepted by the Jewish community, but rejected by the Arabs on the basis that it was inequitable in regards to its proposed land allocation and the fact that it legitimized the Jewish presence in Palestine.

The following year, in 1948, Israel was established as was envisioned by the Zionist movement to be the geographic nexus of the Jewish faith on the remaining of historic Palestine. This created a reality on the ground that was very different from the UN partition plan of 1947, with Israel controlling the majority of the Palestinian land, leaving the Palestinians with only 22% of the land of historic Palestine. While Jewish immigration to Palestine intensified causing nearly 480,000 Jewish immigrants to settle in the country during the years between 1918 and 1948, as shown in Fig. 3,

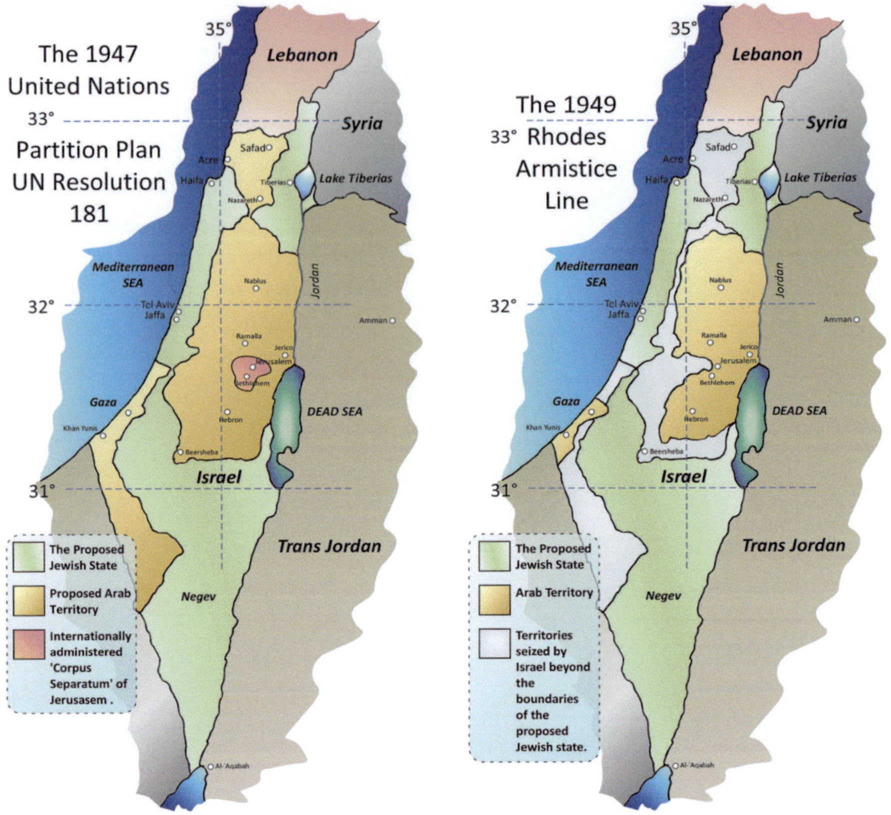

Fig. 2 A map of the UN Partition Plan of 1947 compared to land seized by Zionist militants in 1949. *Source* Illustration rendered by Michael N. Erickson, 2018 based on data from Palestinian Academic Society for the Study of International Affairs (PASSIA)

the initial exodus in 1948 caused over 750,000 Palestinians to leave their homes and villages, as we discussed earlier in the book.

Israel actively renamed these Palestinian villages it seized, depopulated and destroyed in order to erase their memory and rewrite a new chapter in a fabricated history for the region. Israel also attempted to alter the character of the landscape by introducing a new European-looking plantation that gave the land its green appearance, which seemed alien compared to the original character of the indigenous landscape of Palestine (Pappe 2006). Contrary to what the Zionist movement has been trying to propagate about the land of Palestine being uninhabited and effectively empty prior to the founding of the state of Israel, there has been a coherent national Palestinian life that invigorates the Palestinian style and identity well before the existence of the Zionist movement at the end of the nineteenth century (Farsoun 1998). Despite their ties to their land demonstrated by their well-established cultural practices reflecting this unique Palestinian identity, most of the native Palestinian

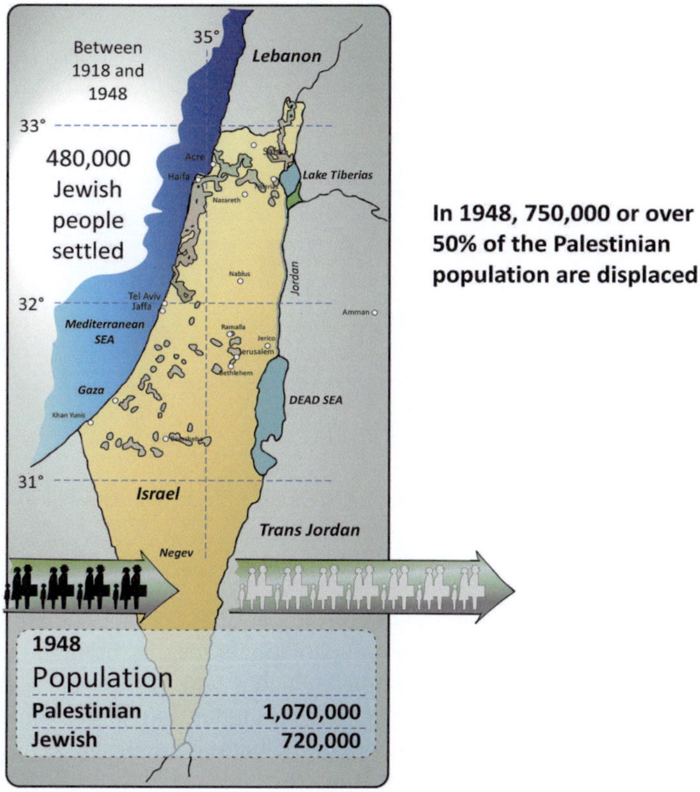

Fig. 3 Jewish immigration and Palestinian displacement between 1918 and 1948. *Source* Illustration rendered by Michael N. Erickson, 2018 (adapted from Visualizing Palestine)

inhabitants of these areas were forcibly removed by Israel to make room for the new Jewish state. As a result, these Palestinians became refugees in their own country. Some of these refugees left to Jordan and Lebanon, while others settled in the West Bank or the Gaza Strip. A new map of the region has been drawn and for the first time the two territories have been separated, as shown in Fig. 4.

The new reality that created and maintained this new map with an ever expanding Israel controlling most of the territories created a physical separation between the confined and much smaller Gaza Strip and West Bank. In this new configuration, the Gaza Strip fell under Egyptian control while the much reduced West Bank fell under Jordanian dominion. It is important to understand that historically the term West Bank was used in reference to all historic Palestine being on the west side of the Jordan River, not just the much smaller West Bank we know today. The two geographic entities, one being under Egypt and the other under Jordan, started to drift away from each other as a natural result of the cultural influence Egypt and Jordan introduced during these years following the 1948 *Nakba*. This was effectively the beginning of

Fig. 4 A map of Palestine in 1948 after the *Nakba*. *Source* Illustration rendered by Michael N. Erickson, 2018 (adapted from Visualizing Palestine)

the cultural gap that started with the establishment of Israel in 1948. However, the need for the Palestinians to achieve a unified national identity brought them closer to each other. The Palestinian refugee issue, which is widely regarded as the largest and most protracted refugee situation in the world today, had direct impacts on the formation of a renewed Palestinian identity. The demolition of the 531 Palestinian villages following the establishment of Israel in 1948 served as the impetus for the Palestinian refugees to keep their memory alive. The collective sense of victimization and nationalism strongly influenced the Palestinian refugee identity that binds them together (Clemens 2007). In this regard, the ongoing opposition and resistance to the designation of refugees served as a strong motivator for uniting the Palestinians and reinforcing their national identity as one people with a just cause. Refugees living in the Gaza Strip, the West Bank and even abroad identified and grouped themselves in imaginative communities based on their original villages and where their originally came from in Palestine. This unique practice of identifying Palestinians (decedents of the *Nakba* survivals) based on their original villages destroyed by Israel during the *Nakba* of 1948 remained alive to this very day.

Palestinians recognize each other and affiliate themselves with the names of their villages in this fashion as a way to associate them with the land and their unique cultural practices and ethnic roots. Despite the fragmented Palestinian identity along lines of geography, experience, and legal status, the collective perception by refugee groups of Israel as a common enemy provided the needed bond to rally them around their shared suffering and a unified cause that they all commonly share and strongly identify with. The "us" versus "them' narrative became pervasive and contributed to establishing a strong sense of nostalgia for their villages and cities and essentially served as a coping mechanism to confront their uncertain difficult circumstances (Clemens 2007). Ironically but intriguingly, the *Nakba*, as haunting and catastrophic as it was, in a way fostered a stronger sense of national identity for the Palestinians who were scattered across the region but sought to maintain and further reinforce their ties to the land they were forced out of. As such, 1948 consecrated the Palestinian *Nakba* which continues to live on until they are granted their inalienable right protected under international law to return to their ancestral homeland.

This map for the most part remained unchanged for the next nineteen years until 1967, when Israel occupied the rest of the country including these two territories during the Six-Day War. Significant Jewish immigration between 1948 and 1967 increased the number of Jewish population to 2.38 million compared to almost 1.1 million Palestinians who were forced to leave during the same time period, causing the Jewish population to exceed the Palestinian population for the first time, as shown in Fig. 5. Despite the new occupation, the economic, social, and cultural ties in all forms that were interrupted with the establishment of Israel and contin-

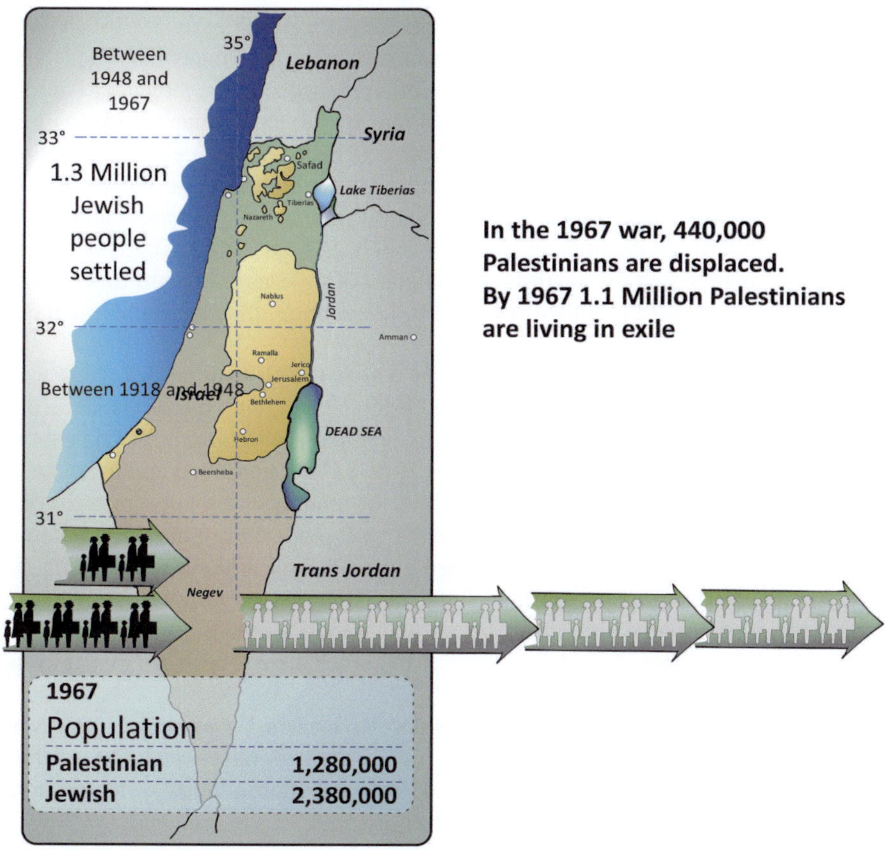

Fig. 5 Jewish immigration and Palestinian displacement between 1948 and 1967. *Source* Illustration rendered by Michael N. Erickson, 2018 (adapted from Visualizing Palestine)

ued during the period between 1948 and 1967 were resumed and largely maintained between the two territories. In some sense, the new occupation brought them even closer together as a one country with a widely understood and unified purpose—getting rid of the malignant occupation with an unshakable commitment and aspiration for self-determination. Economic, social and cultural ties also continued between the Palestinian territories and Israel as the latter employed a significant number of Palestinians from these territories, hoping that improving their economic conditions may cause them to stagnate over independence and statehood aspiration. For the first time since 1967 the Palestinians of the Gaza Strip and West Bank felt that they are one country again sharing one goal despite the obvious geographic separation and the different cultural practices introduced by Egypt and Jordan in the Gaza Strip and the West Bank respectively. The ongoing Palestinian resistance that emanated in the refugee camps of Gaza to the new occupation introduced a new face for national identity and sense of belonging to their cause that was dedicated to ending the occupation and improving their conditions in these camps.

Palestinian refugees in Jordan had a different story in the early 1970s. In 1969 Yasir Arafat, the leader of Fateh at the time, was named the chairman of the PLO. King Hussain of Jordan feared the growing Palestinian presence in the streets of Amman and in the wake of a number of airline hijackings by the Palestinian militia, namely the Popular Front for the Liberation of Palestine (PFLP), he decided to take action. On the morning of September 16th, 1970, King Hussain declared military rule, and on the following day a large-scale military operation began targeting the headquarters of the Palestinian organizations in Zarqa, Sweileh, Salt and Irbid. The result was a raging battle that caused the deaths and expulsion of thousands of Palestinians from Jordan, including the PLO which resettled in Lebanon and later in 1982 moved to Tunisia.

In the wake of this stupendous event in which the Palestinian refugees were displaced once again, the Black September Group (BSG), which took its name as a result of this bloody battle that took place in September, was born out of this defeat in 1970. Made out of recruits from the PFLP, as-Sa'iqa, and other groups close to Fateh, the BSG started to carry out a number of attacks and gained international attention. One of their most publicized attacks took place during the 1972 Summer Olympics in Munich, where they kidnaped eleven Israeli athletes, demanding the release of Palestinian prisoners from Israeli jails. This and the follow-up failed rescue attempt resulted in the killing of all Israeli hostages, a German police officer, five of the Palestinian attackers and the arrest of the other three who participated in the operation.

The bodies of the five Palestinian attackers arrived in Libya where they were welcomed as heroes. The other three Palestinian attackers were freed as a result of the hijacking of Lufthansa Flight 615, and received a tumultuous welcome upon their arrival in Libya with an unprecedented press conference broadcast worldwide of their account of the operation and explanation of their demands which were never met. Although failed to accomplish its demand of releasing the prisoners, this daring operation was successful in putting the Palestinian cause on the international stage. On October 28th, 1974, the PLO was recognized as the sole legitimate representative of the Palestinian people and on November 13th, 1974, its leader Arafat was given a rare opportunity to address the United Nations, where he delivered one of his most recognizable statements, "I have come bearing an olive branch and a freedom fighter's gun. Do not let the olive branch fall from my hand" (CNN Library 2017). Later that year, the UN supported the Palestinian right to statehood and recognized the PLO as an observer status, giving it a formal place at the UN. Palestinians welcomed and viewed these events as part of their national resistance. These events ipso facto had a uniting effect on the Palestinian communities most of whom were refugees from 1948 and 1967 living in the occupied territories or exile.

The sustained occupation of Palestine made it possible for Israel to plant and fuel the seed of division between the Palestinians living in the Gaza Strip and those living in the West Bank. Israel started building illegal settlements in these territories, accelerating its judaisation project of Jerusalem and began to formally adopt separation policies to segregate the two territories from one another. Israel deliberately promoted different policies and even educational and cultural systems in the two territories to maintain and reinforce the physical separation and expand it to include psychological and cultural dimensions. For example, Egyptian curriculum is used in Gaza, while Jordanian curriculum is used in the West Bank. Furthermore, Israel has always planned for the Gaza Strip to be the future Palestinian state, which was demonstrated as the result of the peace process as will be discussed later. With this new reality on the ground, the Palestinian communities living in both territories started to exhibit unique cultural and social differences and display their own distinct identities as a result of decades of separation. This divide became more evident and pronounced with time.

2 Late Twentieth Century and First Palestinian Intifada

In 1987, the first Palestinian uprising (known as the Intifada) started in the Gaza Strip and spilled over into the West Bank and the rest of Palestine despite the geographic separation between them. Although the geographic separation was a reality on the ground that prevented people from moving freely between the two territories, the Palestinians felt as if they were one country united during the Intifada as a result of their united struggle against occupation and oppression. Thus, the Intifada was a uniting factor that brought the two territories closer to each other than ever, with a strong sense of place shaping and cementing an unbreakable Palestinian identity that is closely intertwined to their common refugee experience.

Where opportunities and decent life are limited in these refugee camps, social status among Palestinians is obtained by taking part in the struggle and demonstrations against the occupation. Facing arrest, deportation, injury and even death is looked at as a noble and honorable consequence that was normalized among the majority of the society (Elbedour et al. 1997). This new sense of pride and "status" gained from scarifying oneself to protect Palestine, which often referred to as the "mother," dominated and bonded the Palestinian communities that were strongly rallied behind this narrative. The news about the Intifada traveled from south of the country in Rafah all the way to the north in Naqoura with the speed of light. Images of martyrs were displayed in the streets of both territories as symbols of resistance to commemorate their struggle (Peteet 1995).

A prevailing sense of unity and Palestinian life was emerging to preserve the Palestinian memory. Various objects of memories have been collected and maintained to keep the Palestinian memory alive, such as house keys, newspapers, and maps to name a few. One such potent object that dominated the streets of Gaza and the West Bank and became an iconic feature of the Intifada was a golden map of Palestine (showing pre-partition Palestine) that was so pervasive when I was growing up in the camp. I still recall this remarkable and iconic golden map which was printed and painted on walls of the camps, worn as necklaces and on key chains, and frequently referred to as the Palestinian homeland. It reinforced the sense of belonging and fueled Palestinian interest and aspiration to return back to their ancestral homeland. The country was united from the river to the sea. This story, however, has been interdicted and mutilated by well-orchestrated and willfully deceptive rubric in the American and Israeli mainstream media and public discourses in an attempt to conceal or deny it altogether. According to Said, 2003, "the great narrative of Palestinian life has not yet been achieved any more than the fulfillment of its logical national aspiration has been fulfilled … Despite all this, a Palestinian nation in exile has emerged with an unmistakable outline and personality of its own." However, this changed little of the map that was drawn in 1948 in the wake of the establishment of the State of Israel and was reinforced in 1967 with the Six-Day War.

During the first Intifada, an unprecedented and hugely intimidating military presence is usually used to quell and crush any mass protests. To use military dominance with such belligerence is an indication of lack of concern about Palestinian human rights. Certainly it is the case that this well-documented abrogation of human rights is also indicative of the constant overall sulfurous Israeli position towards the Palestinians as a nation, let alone their independence. When confronted with these atrocities it is committing against the Palestinians, Israel often uses the anti-semitism threat to deter anyone from criticizing its practices against the Palestinians who bear the brunt of it. Such sweeping statements and accusations often lends Israel a status of higher impunity that must be challenged and intellectually dismantled, especially if they are used to justify or cover atrocious human rights violations and vast ruin (Said 2003).

Edward Said astutely contends that "Israeli policies as well as the meditation of many Western liberals who feel a collective guilt for Christian and European anti-semitism nevertheless must be able to disentangle the past from the present to make very clear distinctions between respecting the appalling past and justices heaped on Jews historically on the one hand and on the other simply excusing what the state of the Jews has done to the entire Palestinian people ever since. There is simply no earthy or divine dispensation that excuse a state or a people for wreaking havoc over another while pleading through the travails of its own past as an excuse." He adds "I would always want to say that as a Palestinian I should make my case to defenders of Israel on a basis of taking full account of making due acknowledgement of the history of discrimination against Jews that kind of policy should never ever be visited on anyone for any reason at all least of all by Jews in the name of their safety and independence" (Said 2003).

It is pivotal to understand that Palestinians are not a homogeneous group of people, however. They exhibit many different and distinct cultural practices. For that reason, among others, many of the refuges opted to resist assimilation in the new environment that the camp brought to their lives and as such continued to cling onto their original cultural heritage that their families had prior to their dislocation from their villages in 1948. This aspect alone was influential in tying the refugees to their land and the overall Palestinian social fabric. With oral narrative and diverse intellectual heritage, significant and dramatic collective suffering have been preserved along with national life that tends to outpour and outlive them (Said 2003). It is amazing how Palestinians feel, on a very individual level, the duty to narrate their story of suffering and dispossession to others in archeological sequence as part of their identity and who they really are. This dazzlingly widespread commitment to preserving the Palestinian memory is what sustained the continuity of Palestinian life and kept Palestine alive and well.

Politically, there are two major political movements in Palestine—Fateh and Hamas. These two movements have been at odds with each other since the establishment of Hamas in 1987. Hamas was established in Gaza and expanded into the West Bank challenging its rival, the Fateh party. The increasing support for Hamas in the territories became much stronger with time and created strong opposition to Fateh and the Palestinian Authority, which in turn introduced a new political reality that contributed to the increasing polarization of the already divided Palestinian society in these two territories. For the most part, the majority of Hamas supporters are in the Gaza Strip, while the majority of Fateh affiliates are in the West Bank.

On December 13, 1988, Arafat, the leader of the PLO, recognized Israel and declared a Palestinian state, in a speech at the United Nations (Cody 1988). Israel, however, did not reciprocate and instead continued to deny the right of the Palestinians to exist as a state, maintaining its de facto military control of the occupied territories. In 1991, new and harsh restrictions on the freedom of movement between the two territories were introduced and enforced by Israel making it almost impossible for people to travel between the two territories. A new era of Israeli domination over all aspects of the Palestinian life is about to begin with serious diplomatic attempts to resolve the conflict underway. Although the early 1990s witnessed years of peace talks in an attempt to resolve the ongoing Israeli–Palestinian conflict, it also brought with it resentment and widespread Palestinian armed resistance in response to the ongoing Israeli policies of segregation and military domination. These diplomatic aspects and the Palestinian-Israeli peace talks will be examined in more detail in the following chapters.

Since the initial exodus in 1948, the policy of displacement and deportation continued as a means to quell resistance. In an attempt to shuffle the Intifada, Israel in 1992 carried out the largest deportation of 425 Palestinians who were exiled from Israel, the occupied West Bank and Gaza and sent to Marj al-Zuhur in south Lebanon for one year. This *en mass* deportation, along with arrests and imprisonments, had serious implications for the Intifada which started to slowly dwindle. Although the first Palestinian Intifada was a spark that shattered and eliminated all geographic boundaries and united the people of Palestine behind one mission, the subsequent years following the signing of the Oslo I Peace agreement by the Palestine Liberation Organization (PLO) witnessed the beginning of a much more dangerous divide that eroded the Palestinian relationship and continued to separate the two territories until today. Fateh and its followers were supportive of the 1993 deal with Israel, but Hamas was completely against it, which added more fuel to the Palestinian divide. This divide manifested itself in terms of political affiliation at the beginning but extended to almost every aspect of the Palestinian life. The failure of the PLO to meaningfully engage with almost half of its Palestinian population that opposed it by taking the side of Hamas on the issue is one of the most critical aspects that eventually led to the failure of these negotiations as they lacked adequate support on the local level and resulted in a situation that is less than conducive for the two-state solution to materialize. This move was also perceived as a sellout of the Palestinian cause in an attempt to marginalize a fairly large sector of the Palestinian society who was not involved, consulted or engaged in any meaningful way in these diplomatic efforts, and consequently the Palestinian fabric started to crack. The map of segregated Palestine remained the same and even the many peace talks and agreements between the Israelis and Palestinians failed to change this map. Instead these agreements made it easier for Israel to tighten its grip on the territories and legitimize the status quo while fueling the divide within the Palestinian house.

The 1993 Accord, which was signed in secret in Oslo and later was announced with the hand shake between Yassir Arafat and Yitzhak Rabin on the Whitehouse Lawn, resulted in the establishment of the Palestinian National Authority (PNA), as a self-governing interim administration in the Palestinian territories (the Gaza Strip and West Bank). The significance of the Oslo Accord lies in the fact that it regarded the Gaza Strip and the West Bank as a single territorial unit. As such, these talks also considered building a tunnel to connect the two inseparable territories. Despite the Oslo Accord I in 1993 and II in 1995, Israel's repressive separation policies against the Palestinians were maintained throughout and after the negotiation process. The influence of these policies cannot be overstated as they detrimentally impacted the unity of the two territories that are physically separated by Israel.

In 1994 and as a result of the 1993 Oslo I Accord, the Palestinians were given limited self-rule in the West Bank and Gaza and Arafat was given the chance to return to Gaza for the first time in twenty seven years. In the wake of the Gaza-Jericho Agreement, a follow-up treaty to the Oslo I Accord, people in the territories had renewed hope for a better future, despite the widespread skepticism of these peace talks due in large part to the continuation of Israel's expansive settlement activities. Arafat was welcomed in Gaza as a national leader with a great deal of optimism that he will be able to unite the country and create an independent state for the Palestinians. The Palestinian communities were lined up along the main streets in Gaza in anticipation of the arrival of Arafat who was hailed not only as a hero but also as a key Palestinian liberation symbol. I recall the level of excitement that people had about Arafat's return to Gaza, which was viewed by many as a new chapter and a symbolic first step towards statehood, despite apparent shortcomings of these agreements in terms of both process and outcome. The pervasive rhetoric then was that Arafat and his negotiation team have achieved a great victory and were welcomed as national heroes. There was a clear and overriding sense of normalcy that resounded throughout Palestinian community that their issues have been more or less resolved or are on their way to finding a resolution. The reality however was quite the opposite. Driven by pure human emotions, this irrational exuberance in effect produced a speculative bubble that was quickly dashed with Israel's persistence on imposing a vastly different reality that changed the very geomorphological character of the land in the years to come. The negotiation process and its outcome were both conducted with a great deal of secrecy and resulted in a political blunder that placed the Palestinians and their leadership in a straightjacket and led to a much more difficult situation for the Palestinians, even those who witnessed and naively celebrated these events. Euphoria drove people's expectations up and realism brought these expectations back to earth where the geopolitical situation was less than conducive for achieving peace and reconciliatory measures (Khalidi 2013).

Arafat and the PLO were given a Palestinian authority that was not approved by all Palestinians who felt that these agreements, namely the Oslo Accord, were not representative of the Palestinian hopes and dreams of self-governance. They were rejected by many Palestinians on the basis that they resulted in the recognition of the state of Israel without reciprocal recognition of a Palestinian state. As was explained earlier, many people in Palestine, namely supporters of Hamas, felt that these agreements were not fair to the Palestinians as they made things much worse for them in all aspects, including the socioeconomic and political conditions of the country. In their eyes these agreements gave Israel legitimacy, compromised too much, and created a corrupt government in the Gaza Strip and West Bank without real authority, power or independence from the Israeli occupation which seemed to continue to dictate and control all aspects of the Palestinian life. Instead, the power of this new Palestinian Authority was derivative, can be taken away by Israel at any time, and can only be exercised as long as Israel is permitting it to do so. Any divination meant that Israel can militarily intervene to protect its own interest which according to Israel are related to security and safety to prevent any suicide attacks inside Israel, even if that meant building separation walls and restricting the freedom of movement for the Palestinians. However, the fact that these restrictions were imposed starting in 1991, preceding any suicide attacks on Israel, puts halls in the Israeli argument of security and questions its validity. In addition, the pervasive corruption brought by the Palestinian Authority whose members returned from exile became a daily struggle for the Palestinians living in the occupied territories to cope with. This cost the Palestinian Authority its credibility and legitimacy in the eyes of the Palestinian people who very quickly learned that the Palestinian Authority's job was to effectively safeguard Israel's security as an enforcer of the de facto Israeli occupation. They were viewed by many as a tool to serve Israel's control and interests in the occupied territories.

The ramifications of these agreements on the ground were nothing short of a debacle for the Palestinians who felt its immediate impact so profoundly, namely the

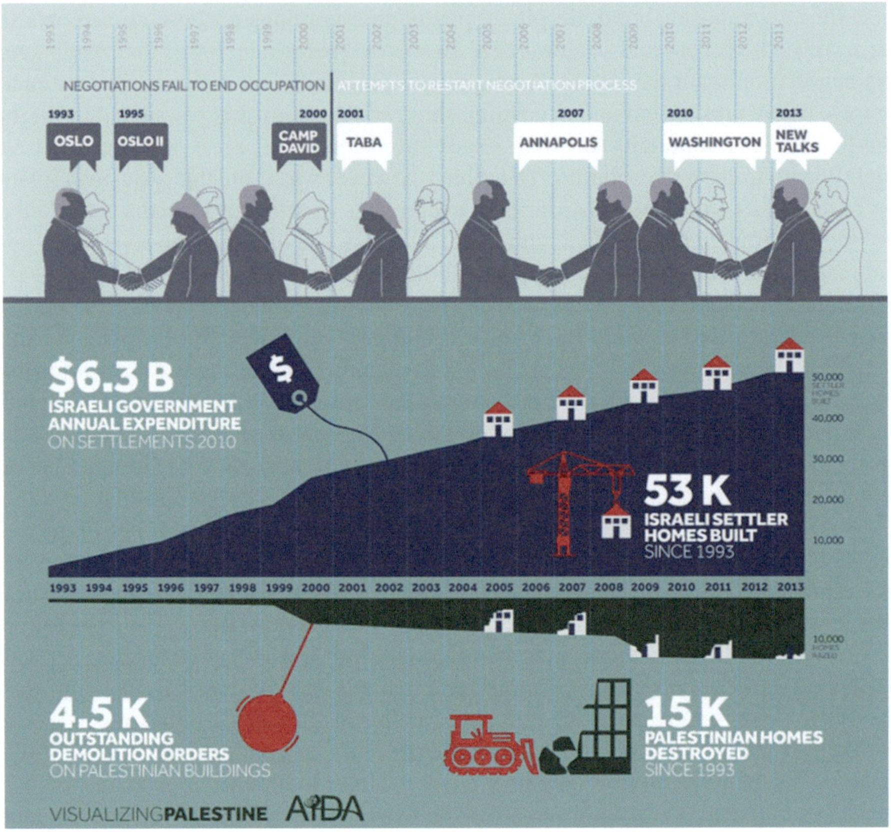

Fig. 6 Twenty years of talks and the impact on Israeli settlements and Palestinian displacement. *Source* Visualizing Palestine (2013)

exponential expansion of Israeli settlements on Palestinian land and the increasing house demolition practices throughout and after these negotiations as shown in Fig. 6. The two territories progressively became more segregated, freedom of movement was severely restricted, and a new wall around Gaza was constructed. The need to travel outside of Gaza is a direct result of the fact that more than one third of the people in Gaza has relatives in the West Bank or Israel. However, the enhanced restrictions on the Palestinians created widespread frustration and anger among the vast majority of the Palestinian society who viewed these agreements as a sellout of the Palestinian cause. Moreover, the political misalignment between the PLO, on one hand, and the Palestinian people on the other, made it difficult for the Palestinians to mend their differences and reconcile their political and geographic disparity. What made things even worse is the fact that the West Bank has always been more supportive, generally speaking, of the direction that the PLO took regarding the Oslo negotiation (and by extension the overall peace process as shown in Fig. 6) of which Gaza was not generally as supportive. This diametrically opposed positions regarding the negotiation process deepened the divide and widened the gap between the two territories.

During these years of negotiations and peace-making effort, Israel implemented many restrictions to deepen the divide and establish two very segregated entities. In 1991, and in conjunction with the ongoing Madrid Peace conference, Israel built the first checkpoint north of the Gaza Strip, known as *Erez*. To tighten its military control and further isolate Gaza, Israel went ahead with its plan to build many more checkpoints around and inside Gaza cutting it off from the rest of the world and making it hard for people to travel even within the Strip. Israel also built a physical barrier around Gaza in 1995 and a wall around and inside the West Bank in 2003 to seal the deal. Gaza is now entering into a new era of segregation, where unless given a special permit, Palestinians in Gaza cannot leave for any reason. Families and relatives became separated along these political and geographic divides. The design and deliberate location of the Israeli wall that sneaks through the Palestinian communities separating people from their families and farmers from their land is meant to effectively plot out the existence of any future Palestinian state (Said 2003). Documenting this harsh reality imposed on the Palestinian communities, Amira Hass explains that "A person could travel the length and breadth of the West Bank without ever knowing—not only the names of the villages and cities whose lands were confiscated in order to build the Jewish settlements and neighborhoods, but even the fact that they exist. Most of their names cannot be found on the road signs. And from a distance, the calls of the muezzins and the streets empty of people (after all, there is nothing to go out for) seem like an aesthetic decoration. A Jew traveling on the almost empty roads of the West Bank would think that there no longer are any Arabs: They do not travel on the wide roads used by the Jews" (Hass 2003).

This also meant that Palestinians in both the Gaza Strip and West Bank cannot visit each other without going through a long and tedious process of approval from the Israeli government, which is designed in a very cynical way to make people give up. This process often results in denial of entry due to "security reasons" for most people in Gaza. There are many Israeli restrictions that prevent those who carry an ID (*Hawia*) issued in the West Bank to visit or live in Gaza and vice-a-versa. This is all based on where a person is registered, which determines where he/she can stay for the rest of their life and freedom of movement from one territory to another. Students from Gaza were prevented from getting permission to study in the West Bank and were further jailed if caught noncompliant. Even students who are admitted to U.S. universities and only seek to get a U.S. visa are not allowed to interview in the U.S. embassy in Tel Aviv or the U.S. consulate in Jerusalem without obtaining a permission from the Israeli government, which is usually not granted on the basis of alleged security. Similarly, families and friends in these territories cannot visit each other without a tedious process of permitting which also usually results in a denial. Although the two territories are separated by a distance of only seventy miles, the geographic separation is compounded by these Israeli restrictions imposed on travel, which all made it feel like a much longer and further distance in reality for those who experience separation and isolation. A Palestinian traveler from Gaza has to go to Egypt, Jordan, and then to the West Bank in painstaking journey through four different countries and territories. Parents in Gaza used to say their goodbyes to their sons and daughters seeking education or visiting relatives in the West Bank as if they were never going to see them again due to these restrictions which sometimes meant students cannot return back home for an extended period of time until they can actually finish their degrees. Students who leave to visit families home may risk not being able to go back to their studies, in many cases costing them their education and future.

The ramification of the continuation and perpetuation of these policies of segregation on the Palestinian society are widely felt by many who experienced it on a daily basis. The political boundaries drawn by the current map represents a much wider divide in terms of national identity and political affiliation. With decades of physical separation, this political, ideological and cultural gap manifested itself in distinct social and cultural differences between the Palestinian communities living in these two territories. The fact that the West Bank is run by the PLO government, while Gaza is run by the Hamas government created two politically misaligned, sometimes antithetical entities. Hamas government opposed the peace agreements with Israel and that position was for the most part held by the majority of people in Gaza who resented these agreements. On the contrary, the West Bank is much more receptive of the PLO's position regarding previous peace agreements, although people on the ground felt little difference in their lives as a result of these agreements. As a result, sentiments of disagreements with these previous peace talks started to penetrate the West Bank as well. Nonetheless, this political polarization significantly contributed to the ever-growing divide between the two territories.

The geographic vicinity of Gaza to Egypt and the West Bank to Jordan also colored and complimented their diverse culture and gave each a unique social and cultural identity. Although very different, Gaza is known to have somewhat similar social norms and practices to Egypt which influenced it as a result of its close geographic location, but also the fact that Egypt was at some point in charge of the Strip. Many people in Gaza have families residing in Egypt and hold family ties with Egyptians who also became more sympathetic to their cause. Similarly, the West Bank society in a way resembles that of Jordan, due to its vicinity and the fact that Jordan administrated it at some point. We can also see similar family ties between the Palestinians of the West Bank and Jordanians.

Although the conflict between Hamas and Fateh was rooted in the Oslo agreements, it intensified in the mid-1990s with the new wave of suicide bumping attacks inside Israel carried out by Hamas who opposed these Fateh-sponsored agreements with Israel. These attacks were in response to the 1994 Ibrahimi Mosque massacre in Hebron by Baruch Goldstein, an Israeli settler who shot and killed twenty nine Palestinian worshipers. Due to mounting pressure on the PLO and Fateh from Israel and the U.S., the Palestinian Authority led by Fateh denounced Hamas attacks and began to arrest and jail its members in the territories. This turn of events took the internal Palestinian conflict to a whole new level of escalation and tore the Palestinian community along those two movement lines. The acquiescence of the PLO to Israel's demand to imprison members of Hamas signaled a seismic shift that rocked the Palestinian unity and sent shockwaves into Palestinian communities until this day. Supporters of Hamas and Fateh, who are diametrically opposed to each other in regards to the process and outcome of the peace talks, were put on a collision course with these arrests that continued to this day and created a massive rift in the Palestinian house. Although their lives as well as their destiny are tied together, their ideological differences seem to be irreconcilable, at least for the time being.

3 Modern Day Palestine

The land of Palestine went through a series of portioning plans, geopolitical events, and confiscation that severely altered its size, shape and even its geographic and demographic characters. The increasing Jewish immigration between 1967 and 2008 of 1.8 million newcomers drastically changed the demographic characteristics of the region, as shown in Fig. 7. Today both the Gaza Strip and West Bank are part of what's known as the Occupied Palestinian Territories. The demographic composition of Gaza is very unique as it is made of Palestinians who are native to Gaza but also many other refugees who were displaced and forced out of their homes in the wake of the establishment of Israel in 1948. In addition to the local Gazans (in Arabic *Moatineen*), Gaza houses eight refugee camps. Refugees (or *Laje'en* in Arabic) make up the predominant part of the Palestinian population in the Gaza Strip, which accounts for the overwhelming majority of more than 70% of the current population according to the UNRWA. The term "refugee" has been used in Palestine in a very unique way to affirm the right of a group of Palestinians (the refugees) to return to their homelands and that these rights are yet to be met as long as the term exists. Years of conflict and blockade exacerbated the divide between the two territories and left eighty percent of the population in Gaza completely dependent on international assistance for survival. With the local economy on the verge of collapse, the average unemployment rate is well over 40% signaling dire economic conditions.

Due to its high number of refugees, Gaza represents a unique compilation of demographic and cultural heritage that can only be found in this tiny enclave. With the majority of the Gaza Strip's population made of Palestinian refugees from all over Palestine, Gaza represents a much more diverse cultural identity than the West Bank, making the Strip a microcosm of cultural diversity akin to the whole of Palestine. In Gaza, you can find the vast majority of Palestinian dishes, cultural practices, and even dialects. The fact that Gaza exhibits a very interesting mix of cultures made it a much better representation of the rich pre-1948 Palestinian cultural diversity, despite its small land area. As such, Gaza is commonly thought to be a more complete and accurate representation of pre-1948 Palestine in terms of its demographic characteristics as well as cultural practices. This is due in part to the fact that these refugees did not assimilate to other local cultures and instead continued to adopt their ancestral cultural practices originally found in their villages they were forced out of in 1948, but continue to identify with. The West Bank, on the other hand, is comprised of

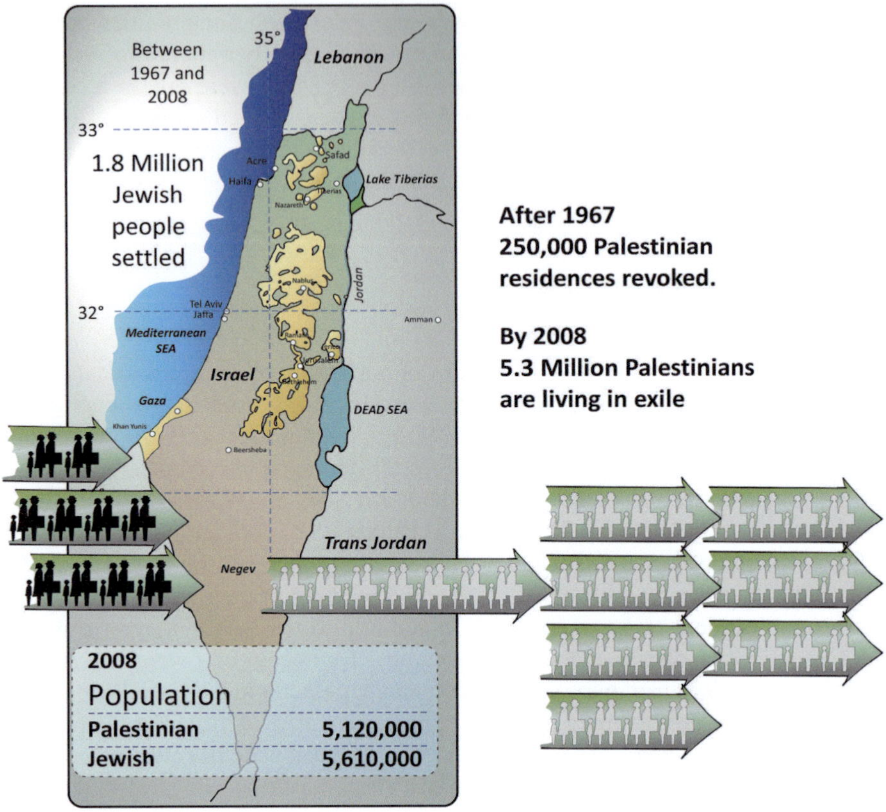

Fig. 7 Jewish immigration and Palestinian displacement between 1967 and 2008. *Source* Illustration rendered by Michael N. Erickson, 2018 (adapted from Visualizing Palestine)

smaller villages and towns that maintained for the most part homogeneous cultural identity heavily influenced by neighboring Jordan.

In 1997, Sheik Ahmed Yasin, leader of Hamas, was released from Israeli prison as part of a deal with Jordan in the wake of the failed assassination attempt of Khaled Mashal by the Israeli Mossad inside Jordan. The release of Sheik Yasin (who was later assassinated by Israel in 2004) in this token deal between Israel and Jordan paved the way for the reconciliation effort in order to build trust between the two Palestinian factions. This united front emerged even stronger with the eruption of the Second Intifada in 2002 as a result of the obvious failure of the peace process to realize any of its promises towards establishing self-determination and addressing the issue of Palestinian refugees, or at the very least improving the lives of the Palestinians who live under miserable conditions inside occupied Palestine or outside under authoritarian Arab regimes. It all started in September 2002, when the right-wing Israeli opposition leader, Ariel Sharon, visited Temple Mount in Jerusalem causing widespread condemnation and anger. This stupendous visit and the continued

Israeli policy of segregation and house demolition caused a wave of frustration that culminated in the eruption of the Second Intifada in 2002. The Second Intifada personified this mounting frustration with the conditions under which the Palestinian population has living the deterioration of which is often attributed to the failure of the aforementioned peace process. For many of those Palestinian refugees, the Second Intifada also meant to send a clear message to Israel that they cannot have a free field to expand their settlements while denying the Palestinians their freedom and the right to self-determination.

In response to the mounting violence, Israel escalated its military campaigns by bombing Gaza from the land, sea and air, killing and injuring Palestinian civilians and taking the conflict to new heights with the use of heavy weaponry against unarmed civilians. Israel utilized more checkpoints inside the Gaza Strip to divide it into smaller parts which severely restricted the movement of Palestinians within the tiny Strip. The movement of patients, students and other sectors of the society were severely cribbed as a direct result. The sight of hundreds of cars waiting at Israeli checkpoints became the hallmark of this time period and signaled a clear *prima facie* evidence of the failure of the peace process. Under these harsh conditions, Palestinians' ability to work, study, or seek medical treatment is all scandalously curtailed by statuary laws designed for the purpose of confining them and exerting pressure. These policies are designed to tear families apart and keep them separated from each other. The utter inequality between the stateless and dispossessed Palestinian indigenous population and the newcomers is evident in the conscious willful abuse of human rights directed against a downtrodden dispossessed people. Frequent Israeli incursions of Gaza often result in house demolition, detentions, and killing of Palestinian civilians. These infringements and brutalizing assaults pervasively take place on a frequent basis as a matter of practice. As Edward Said puts it, neither Indians demonstrating against the British nor the blacks in South Africa fighting against apartheid had to reckon with tanks, F-16s, and missiles fired into civilian areas with no defenses or arms the way Palestinians do (Said 2003). As harsh and brutal they are, these events and conditions eased the tension between Hamas and Fateh and brought them closer to each other, giving the Palestinians an opportunity to stand united again. With the new Intifada in full swing, armed Palestinian groups started to coordinate combined attacks against Israel. This was the beginning of the end for the so called "peace process" between Israel and the PLO.

In 2004, Arafat died after a long and grueling Israeli siege of his Ramallah headquarters in Al-muqataa. His death was surrounded by mystery and for a little while united the Palestinians as one nation mourning the death of their beloved leader whose iconography transcended the political discourse, not only as a political leader but also as a uniting figure. Both Hamas and Fateh, as well as other Palestinian factions, were all shocked by the sudden and mysterious illness that led to the death of one of the most recognizable symbols in the Palestinian struggle for freedom and statehood. His body was flown to Ramallah were it received a hero's welcome by all Palestine and was later laid to rest. His death marked a new era in the Palestinian-Israeli conflict. He was succeeded by current Palestinian President Mahmoud Abbas who does not enjoy the same level of support from the Palestinians that Arafat had. Although his death briefly brought all Palestinian factions closer together, the divide between the parties was renewed shortly after. Hamas boycotted the Palestinian election that resulted in the wining of Fateh leader Mahmoud Abbas as the new Palestinian President on the basis of their disapproval of the Oslo Accord that led to the establishment of the Palestinian Authority in the first place. Hamas and Islamic Jihad joined forces against the Israeli occupation of the territories and intensified their attacks.

In 2005, President Abbas and Israeli Prime Minister Ariel Sharon formally agreed to a mutual cessation of violence marking the end of the second Intifada in the Sharm al-Sheikh Summit held in Egypt. The Palestinians at this point lost faith in negotiations with Israel due to the fact that their conditions went from bad to worse during these years of negotiations, while sporadic armed conflicts continued in the Gaza Strip.

Due to the ongoing attacks, Israel realized the heavy cost of maintaining its military presence in Gaza. The Israeli military and agricultural settlements as well as military posts and checkpoints in Gaza were frequently attacked, which effectively caused Israel to unilaterally withdraw from the Strip in 2005. This gave rise to the idea that the peace process has completely failed and was abandoned by both sides. The notion of an independent Palestinian state seemed insurmountable and the hard reality that the peace process has collapsed began to permeate both the Palestinian and Israeli societies alike. This unilateral disengagement reinforced the separation between the Gaza Strip and West Bank on one hand, and paved the way for one of the most brutal sieges in human history, on the other.

On January 25, 2006, Hamas ran in the Palestinian election as the "Change and Reform" list, and won a substantial majority of the Palestinian Legislative Council seats, pushing its rival Fateh out of power and marking a new chapter in the internal Palestinian political discourse. With its election to role Gaza, Hamas sought to form a coalition government but all factions refused to participate. In addition, Fateh did not submit to an orderly transfer of power. The PNA and Fateh, led by Mahmoud Abbas in the West Bank, ordered its supporters in Gaza to rebel against the newly elected Hamas government by encouraging them to boycott the government and go on a public sector workers strike. Many of Fateh supporters complied and stayed home while their salaries continued to be paid by the Abbas's government in the West Bank. Those who did not abide were punished by the Abbas's government by deliberately cutting off their salaries although they continued to go to work and provide services. This was viewed by many Palestinians in Gaza as unwarranted collective punishment on the people of Gaza and as a result pushed the two territories further away from each other. The impact that this decision had was tremendous in worsening an already divisive relationship and tearing the Palestinian fabric apart.

In 2007, their disagreement came to a head and the two parties clashed in the streets of Gaza in what is known as the battle of Gaza which lasted five days. After this violent intra-Palestinian battle that lasted for a few days and resulted in the killing of hundreds of Palestinians, Hamas gained control of the Gaza Strip and pushed the remainder of Fateh-allied officials, leader and security members to the West Bank, widening the already existing separation gap between the two territories. Both Hamas and Fateh exchanged blame for the situation created in the occupied territories. Hamas blamed Fateh for collaborating with Israel and the U.S. to impose a siege on Gaza and attempting to topple Hamas, while Fateh blamed Hamas for the halt of the international donor funding. Israel immediately imposed a naval and land blockade on Gaza that continues until today preventing the movement of people, goods, and medicine from entering or leaving Gaza, turning the tiny enclave into what's widely known as the largest open-air person in the world. Israel closed all passenger, industrial and commercial crossings surrounding Gaza and further stopped the transfer of any taxes collected on behalf of the PNA to Gaza as part of this siege.

Although Israel unilaterally disengaged from Gaza in 2005 relocating its illegal settlements and military posts outside of Gaza's borders, Gaza is still considered to be occupied by Israel which controls its borders, airspace, and waterways. Gaza has seven land crossing points- some are dedicated for commercial purposes while others are dedicated for passengers' movement, six of which are controlled by Israel while Egypt controls the Rafah border crossing mainly dedicated to passengers leaving and entering the Strip. Israel controls externally what gets in and what gets out of Gaza, when, how, and how much, directly impacting the lives of the Palestinians residing in it. By maintaining its tight control over its water supply, telecommunications, electricity and other essential services, Israel effectively holds the Gazan population captive, which is considered to be a major market for Israeli goods as it has become completely economically dependent on Israel.

Gaza is surrounded by barbed wires and has been under illegal siege by both Israel and Egypt since the election of Hamas in 2006. Although Gaza has been under siege for many years prior to 2006 with its borders opening intermittently and unreliably for passengers, the current siege meant that these entry points are completely closed off. This siege also meant that people cannot leave or enter Gaza freely restricting their freedom of movement, which is a major violation of international law. In this regard, people cannot freely move out or into Gaza, which tends to hinder many people's ability to travel outside of the territory for education, business or medical treatment. Many families have not been able to see or visit each other because of the restrictions on travel imposed by this illegal Israeli-Egyptian blockade. The local economy is significantly impacted and can be said to be for the most part crippled by the siege. Gaza is not only economically disadvantaged and cut off from the rest of the world, but it also seems to be completely marginalized and abandoned. With the deterioration of the situation in Gaza on all fronts, people in the territory have a general feeling of resentment towards the West Bank and its government which is viewed as a deserter and an active participant in implementing this cribbing siege. As a result, it is no surprise that the population of Gaza is frustrated and has little hope in peace.

With the continued restrictions and the suffocating Israeli-Egyptian siege on Gaza, Israel launched several wars on the Strip in 2008/2009, 2012, and most recently in 2014, with various excuses, ranging from stopping the rocket attacks and destroying the tunnel network, to the complete removal of Hamas. Although Israel miserably failed to achieve any of these goals, these wars resulted in documented wars crimes, causing many thousands of civilian deaths among the Palestinians and making an already dire situation even worse. In June 2014, and with mounting pressure from the

Palestinian and Arab streets, Fateh and Hamas agreed to form a unity government. However. Israel prevented government officials from traveling between Gaza and the West Bank to ensure the failure of this newly formed government and to maintain the status quo of Palestinian segregation and Israeli domination over the territories.

In July, a Palestinian teen was kidnaped and burned alive by Jewish settlers in the West Bank and three Israeli teens were kidnaped and killed. Israel blamed Hamas although the latter denied responsibility. Nonetheless, Israel launched its most devastating military campaign against the besieged Gaza Strip to foil this unity government and prevent it from succeeding. This large-scale war lasted for 51 days resulting in over 2000 Palestinian civilian deaths and the destruction of most of Gaza's infrastructure. Some estimates indicate that it would take at least 20 years to rebuild Gaza. Israel's action in Gaza was once again characterized as nothing short of war crimes. When the dust settled, Gaza emerged as victorious politically and militarily, while Israel appeared unsuccessful in accomplishing its goals despite being the Goliath in the war. This war, however, was used by Israel as a tool to enhance the division between Gaza, which has been fighting for its survival, and the West Bank, which did not seem to be impacted much by this war, signifying the pronounced disconnect between the two territories. The failure of this unity government personifies in the minds of the Palestinians that the two-state solution may have become an illusion rather than a practical reality. Israel since then intensified its separation policies and activities, utilizing its expanded settlements in the West Bank as physical barriers to prevent and encircle the Palestinian communities and cities from expanding or growing. This could potentially create smaller geographic entities that are completely segregated from one another, ending any hope of a practical two-state solution on the 1967 territories.

In 2016, Shimon Peres, Former President of Israel, died and Palestinian President Mahmoud Abbas visited his funeral. This triggered a wave of anger among Palestinians who viewed his visit as sympathetic to a war criminal and ipso facto a sellout. In the wake of his visit, Fateh supporters started to display a pronounced divide even among their party regarding Abbas's credibility and his attitudes towards Israel. This negative perception also extended to the entire peace process that Abbas was advocating for which also lost legitimacy and credibility in the eyes of the Palestinians. The vast majority of the Palestinians view the ongoing segregation and isolation of the Palestinian communities in the West Bank as a direct result of this peace process that produced the PNA. Figure 8 shows the extent of fragmentation of the Palestinian communities in the West Bank as a result of Israeli settlement, exclusive road network, and the Israeli wall that sliced through the territory. It also depicts the magnitude of the Palestinian refugee problem which has yet to be addressed in any of these diplomatic engagements that the PLO undertook under the auspices of the U.S.

The fierce internal clashes heightened calls for the urgent need to bring this internal conflict between the two factions to an end. Various and futile national reconciliation efforts took place at many stages of this national divide to create national unity. A Saudi-sponsored initiative brought the parties together in an attempt to negotiate national reconciliation in Mecca (JMCC 2009). Fateh and Hamas representatives

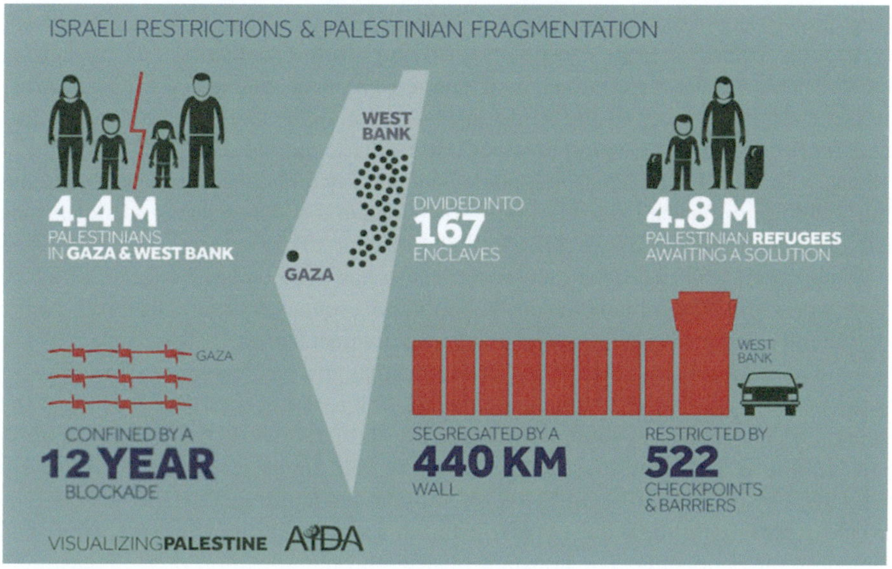

Fig. 8 Israel's restrictions and Palestinian fragmentation. *Source* Modified from Visualizing Palestine (2013)

signed the Mecca agreement in 2007 that called for an end to the internal Palestinian violence and a formation of a national unity government. This brief unity did not last for long as it was interrupted by the Hamas takeover of the Gaza Strip. Fateh as a result formed an emergency cabinet in the West Back that excluded Hamas, dismissing the unity government. This rift that can only be described as a self-inflected wound effectively created two governments vying for recognition and legitimacy—one in Gaza and one in Ramallah (JMCC 2009). Critical of these governments and their insidious role in exacerbating the divide and by extension their suffering, the Palestinians now raise a whole raft of questions about the legitimacy of both governments and their ability, or lack thereof, to serve the interests of their people. This bitter divide has been the perfect storm for the Palestinian communities who are confronted with multiple layers of challenges, internally and externally.

Other national reconciliation attempts were conducted to contain the situation and end the bitter divide between the West Bank and Gaza Strip governments, including the Yemeni, Arab League and Egyptian initiatives. These efforts consistently failed to yield any agreements that sustained implementation. The only brief incidents these factions found themselves aligned was to confront Israel's aggression during the wars waged against Gaza. The ideological divide, the lack of consensus fueled by the irreconcilable differences pertaining to the peace process, and Israel's segregation policies all contributed to the sustainment of the polarization among the Palestinians and the demise of the unity government. According to Khalidi 2013, "the hard line taken by these powers [Israel and the U.S.] exacerbated, as it was meant to, the already deep inter-Palestinian division between Fateh and Hamas. This division was abetted by Israelis-American-European policies of supporting one faction and boycotting another. This should be obvious, as a divided Palestine polity could not possibly make or implement any of the hard long-term decisions about war and peace that meaningful negotiations would involve." He further expounds that "the exclusion of Hamas was among the factors that made such an outcome [successful negotiations] impossible and that call into question how serious the Americans and Europeans actually were about resolving this conflict" (Khalidi 2013).

Since the 1948 *Nakba*, the Palestinian cause has been taking center stage in the Arab world directly influencing its geopolitical events. However, this role began to gradually decline with the Palestinians' involvement in the peace process as the Palestinian issue was assumed to be on the right path to a just resolution. The collapse of the peace process, the Second Intifada, and the recent wars in Gaza brought the Palestinian issue back to square one renewing the Arab world's interest in finding a resolution to this central issue. However, the Palestinian cause has been heavily politicized by many Arab regimes to gain support and legitimacy. Many of the totalitarian Arab regimes utilized the Palestinian issue as a tool to control their people and alienate their opponents. The constant misuse and manipulation of the Palestinian issue by these totalitarian regimes created an adverse perception and association to these regimes and their repressive narrative to control the masses, which in turn contributed to its abandonment. Additionally, in the western public discourses, as is the case in many parts of the Arab world, the Palestinian cause has been associated invidiously with terror, violence and dictatorships. There is a serious contradiction to this superficial association in the ongoing and recognized Palestinian life and struggle for freedom and self-determination that contributes to the narrative of human freedom and highlights this association eccentric tenacity (Said 2003).

The Arab Spring has been one of the most decisive and far reaching event in the Middle East in recent history that changed the overall geopolitical landscape of the region. Millions of Arab youths took to the streets demanding democracy, freedom and human rights, refusing decades of authoritarian regimes, and hoping for positive change and an end to the strangling socio-economic conditions imposed on them for many years. Without delving into detailed historic narrative of the Arab Spring, its causes and outcomes, this far reaching event resulted in momentous, albeit not all positive changes in many of the affected countries, including Egypt, Libya, Tunisia, Yemen, and Syria which has been suffering from Bashar Alassad's brutalization (with support from Russia) to quill the unrest. One cannot underestimate the impact of the Arab Spring on the Middle East peace process. It undoubtedly imposed its own agenda on the overall geopolitical arena not only in regards to the peace process, but also in terms of the shifting of roles of key players in the conflict, including that of Egypt, Turkey and even the United States. The election of Egyptian President Mohammed Morsi, which was a direct result of the Arab Spring and the revolution in Egypt, brought a temporary ease on the movement of the Palestinians in the Gaza Strip. However, in the wake of the military coup that led to the rise of the current Egyptian President Abdel Fattah el-Sisi, a campaign of incitement against the Palestinians was launched accusing them of taking part in the Sinai terror events to provoke chaos and destabilize the new Egyptian government. This has been the pretext for the continued closure of the Rafah border crossing by Egypt and caused the Palestinian-Egyptian relationship to deteriorate. Similarly, the previous Arab support of the Palestinian cause dwindled as these Arab nations have become inundated with their own internal struggles that the Arab Spring triggered. In addition, the U.S. unwavering support and clear partiality towards Israel has become apparent to both the Arabs and Palestinians and as such the U.S. lost credibility to act as a neutral broker of peace between Israel and Palestine. This created a vacuum that Turkey has been unsuccessfully trying to fulfill recently as a key regional player, both on the diplomatic front as well as the refugee crisis. The media preoccupation in the coverage of the Arab Spring, namely the massacres committed throughout the Arab world to quell the unrest, took focus away for the Palestinian issue. This harmed both the Palestinian-Israeli peace process as well as the reconciliation effort to unite the Palestinians, both of which were dwarfed by the magnitude of the regional instability caused by the Arab Spring.

Cognizant of the failure of the peace process, their current divided leadership's ability to make their aspiration for independence a reality, and the increasing pressure imposed on Gaza by the continued crushing siege, Palestinians started to mobilize a different form of non-violent resistance to voice their frustration and refocus the world's attention to their issue. Examples of these peaceful means are the Great March for Return and the Boycott, Divest, and Sanction (BDS) movement. First, thousands of Palestinians amassed at the border with Israel at the start of a prolonged protest, dubbed "The Great March of Return," which was triggered as a response to the Trump's recognition of Jerusalem as the capital of Israel and the anticipated U.S. embassy move to Jerusalem. This is regarded by many to be the largest civil protest that ever happened anywhere in the world in demand of freedom and equal rights rather than supplicants or second class citizens whose rights are utterly violated by the very state that expelled them from their ancestral homeland in the first place. Marking the 70th anniversary of the establishment of Israel, the Great March of Return is a non-violent mass demonstration by the people in the Gaza Strip, the majority of whom are refugees from towns and villages just beyond the edge of the lens across the border. They have been suffering from decades of grinding occupation and a chocking and indiscriminate siege that completely disseminated their local economy. This was a popular organized mass mobilization planned to continue for several weeks to challenge the occupation and the continued policy of engagement imposed on them by Israel, demanding their right to return to their villages and towns as stipulated by the 194 UN Resolution.

These peaceful demonstrations were met with brutal and indiscriminate military response from Israel killing and injuring thousands of unarmed civilians, including children, journalists and paramedics in what has been widely criticized and described as a merciless massacre. The very little and misleading media coverage in the U.S. of the situation portrayed it with a torrent of pro-Israeli narrative and a myriad of manipulative reporting to either justify the killing of Palestinian protestors or to avoid giving a full account of their demands altogether. The pervasive U.S. narrative represented the issue as Israeli soldiers simply defending themselves in the face of

Palestinian aggression directed by Hamas, despite the documented fact that these protesters were unarmed non-violent and the deliberate targeting of these peaceful protesters by Israeli snipers strategically positioned along the border with a shoot-to-kill order. This violent dehumanization of the Palestinians echoes a long history and deep-rooted legacy of manipulating facts in order to discredit and cast aside any Palestinian attempts at demanding their freedom and the world's attention to their just cause regardless of the means used. Highlighting the important role these communicative discourses play in shaping and even corrupting narratives, manipulating facts and creating manufactured reality, Orwell in his book "Politics and the English Language," contends that "political language is designed to make lies sound truthful and murder respectable, and to give an appearance of solidity to pure wind." This assertion is also echoed by Khalidi who, along the same vein, brilliantly ends his book, *Brokers of Deceit*, with this a remarkable conclusion that summarizes the essence of the issue by demonstrating "… how in American political discourse, lies about Palestine are made to be sound truthful; how crimes—against a people and against humanity—are made respectable; and how the pure wind of terms like 'peace process' are given the appearance of solidarity" (Khalidi 2013).

Second, an increasing and growing voice that is getting significant traction is the Boycott, Divestment, Sanctions (BDS) movement. Inspired by the South African anti-apartheid movement, BDS is a Palestinian-led initiative whose goal is to pressure Israel to abide by international law in order for the Palestinian to achieve freedom, justice and equality. To that end, Palestinian-American human rights attorney, Noura Erakat contends that "What we need is to impose sanctions and pressure on Israel to end its apartheid regime" (Erakat 2017). These examples signal a seismic shift in the way Palestinians now can voice their discontent and peacefully demand their rights. The duality and duplicity of the narrative of these activities in the U.S. and Israel's public discourse created a manipulated and inaccurate depiction of these events, their goals, by extension their legitimacy. In many cases, such opposition to Israel is automatically marginalized, demonized and even criminalized in an attempt to cast aside any voices that can exert pressure on Israel to comply with international law. Unless a fundamental change to their current situation occurs on these fronts, Palestinians will continue to seek similar non-violent measures to pursue their inalienable rights and advance their cause.

In a nutshell, the transformation of the land of Palestine and its people, part of whom live within the confines of the Occupied Palestinian Territories or in diaspora, has serious implications for the peace process and the viability of the notion of the two-state solution. Given the current geographic and demographic making of what remains of Palestine, which can only be characterized as non-contagious, segregated enclaves, the two-state solution appears to be fantasy rather than a real possibility. Many Palestinians as well as Israelis caution that this new configuration of the land and the way both peoples interact with it and each other may have resulted in the requiem of the two-state solution, which given the current geopolitical landscape appears to be a phrase at best. The noncontagious nature of these territories remained for the Palestinians and the strict Israeli control and segregation policies makes these areas unsuitable for the establishment and survival of any operational state for the

Palestinians. The implications of such grim reality can only suggest that any future negotiations between the two sides, cognizant of this major limitation on the ground, must look for alternative solutions. Simply implying that a two-state solution can be achieved without major concision from Israel to at least retreat back to the pre-1967 boundary reflects an ignorant view of reality and can only result in a futile and meaningless effort. Israel, whether through negotiations or political and international pressure, must abide by international law to allow for the pre-1967 Palestinian land to be recovered so they can resume negotiations with the Palestinians on issues of self-governance and nation building. The establishment of a viable and independent Palestinian state with its own sovereign boundaries and contagious land can only mean connecting the Gaza Strip to the West Bank to allow for unrestricted freedom of movement between the two territories. This "connection" cannot be just a narrow corridor that goes through Israel and can potentially be closed or even destroyed when things do not go the way Israel wants. The current map of Palestine needs to be redrawn back to the pre-1967 boundaries with a significant expansion of the Palestinian land to allow for a substantial connection and physical contiguity between the Gaza Strip and the West Bank. This can only be accomplished by demolishing the Israeli walls around and within the West Bank and Gaza Strip as well as its illegal settlements, and relinquishing substantial land between the two territories. This is no small change that requires significant diplomatic effort to ensure consensus on, and implementation of, a satisfactory final solution.

In most recent years, a growing alternative that is getting significant traction is the one-state solution. It is being proposed as a more appropriate approach to achieving a workable pre-1948 arrangement, considering the current reality created on the ground. Although a possibility that cannot be ruled out, the practicality and acceptability of the one-state solution is still in question and may appear problematic on many fronts. Either way, a new leadership that can unite the Palestinians is warranted as a first step. A priority must be given to uniting the Palestinian communities under one representative government capable of negotiating and implementing an acceptable and satisfactory deal to ensure just and equitable allocation of physical resources, including land and water necessary to establish and maintain a viable Palestinian state. What is badly needed here is a legitimate force of unification capable of rallying the Palestinians behind a united vision for peace to give peace a chance. It is crucial that the Palestinians are able to patch up their bitter feud in order to represent a united front in future negotiation that allows them independent and legitimate representation of their people. This requires incredible courage, determination and a seismic shift in the Palestinian leadership to transform the current situation not only to a unified Palestinian leadership but most importantly a "Palestinian national consensus on how to proceed" (Khalidi 2013). Khalidi astutely points out that "if the Palestinians do not help themselves, and transform that part of reality which is largely in their power, nothing can begin to change in their situation, nor can anyone be expected to act on their behalf. Self-reliance of this sort is the essential first step, the sine qua non, required to change the pernicious status quo under which the Palestinian people have suffered for so many years" (Khalidi 2013).

Similarly, the Israeli leadership must be willing to take major steps prior to any future negotiations. These steps are not to be seen as preconditions to negotiations, but rather considered as obligations for Israel to fulfill for any meaningful negotiations with the Palestinians to take place. These include denouncing violence, halting all illegal settlements, dismantling the wall, ending the illegal siege on Gaza and abiding by the international law by pulling out of the 1967 occupied territories. Only then will both sides feel that they have a partner in peace and can lead a win-win negotiation towards achieving a just, robust, and lasting settlement. This warrants the applicability of a transformative approach to conflict resolution in order to intercede between the two parties, eliminate hostility, and engender peace for all, which will be discussed in detail in the last chapter of this book.

Chapter 4
Previous Agreements—A Chronicle of Sordid Treachery

The Palestinian Authority belongs to a narrative that doesn't reflect reality on the ground.

—Ilan Pappe

This chapter provides a practical examination to extract lessons learned and illustrate what went wrong in the past agreements and further outlines a new approach that takes into account all aspects of success and avoids elements of failure of these past agreements. It will expose how the conflict has been shaped and mismanaged throughout the past century and right on through the present day. It also provides an analysis of the Israeli-Palestinian peace process, namely the Madrid conference of 1991 and the Oslo Accord of 1993 between the government of Israel and the Palestine Liberation Organization (PLO) and the sustenance of the ensuing talk-flight situation. In tracing the genealogical evolution of the Middle East diplomatic effort, this chapter will pierce key elements and pitfalls of the so-called "peace process" and how it created systematic disadvantage for the Palestinian communities by greatly changing the manifested reality on the ground. This climate of political depiction and obscurity that these diplomatic efforts established will be given special attention to unravel the difference between their stated goals and actual outcomes.

Putting special emphasis on the significance of water, this chapter provides a brief background of the nature of this negotiation process and what it produced. Outlining the pitfalls of these agreements and diplomatic efforts, namely that the Oslo process was overly focused on providing solutions to the problems created by the 1967 Six-Day War, while ignoring the more pressing and catastrophic problems caused by the 1948 Arab-Israeli war, this chapter attempts to draw lessons and conclusions that will be useful in future water negotiations. This chapter will also highlight issues related to the strenuous lack of adherence to their principles by Israel in relation to settlement expansion in the West Bank, severing East

A. Abukhater, *Palestine - Peace by Piece*,
https://doi.org/10.1007/978-3-030-03985-1_4

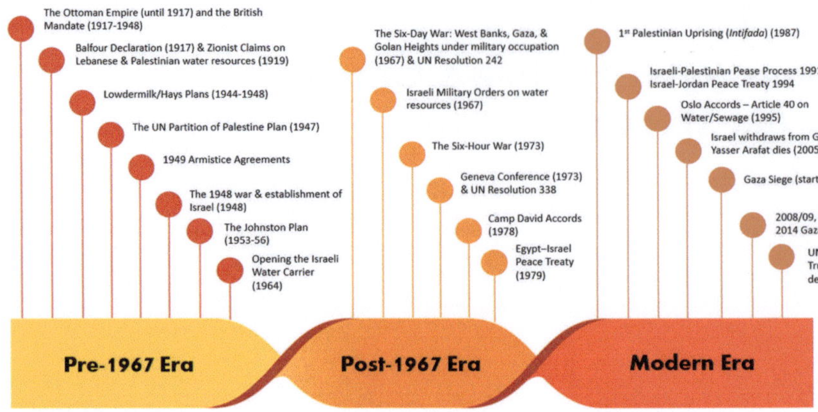

Fig. 1 A timeline demarcating key hydro-political and geopolitical events

Jerusalem from the Palestinian hinterland, the issue of the right of return of Palestinian refugees driven from their homes during the Arab-Israeli wars of 1948 and 1967, the continued blockade and illegal siege of the Gaza Strip, and the failure to grant the Palestinians sovereignty over the occupied territories and their inhabitants as stipulated by these agreements. Special attention will be given to the three wars in Gaza and the current Palestinian uprising, known as the Third Intifada, which is currently in full swing in the Palestinian West Bank and Gaza Strip and is a reflection of the mounting frustration with the peace process and a clear Palestinian rejection of its outcome and the current status quo it produced.

Intermittent negotiation attempts aimed at finding a peaceful settlement and an end to the Israeli-Palestinian conflict is often referred to as the "peace process." This process started as early as the 1970s, where certain Arab countries, such as Egypt and Jordan were able to successfully reach agreeable peace treaties with Israel. These treaties, however, were concerned only with finding mutual basis to independently resolve the conflict between Israel and these individual Arab countries by singling them out in a bilateral fashion one country at a time. As such, these agreements had no bearing on comprehensively resolving the conflict between Israel and Palestine, which is considered one of the most contentious disputes in the world, nor did they provide fair and mutually beneficial treaties for the aforementioned Arab countries involved. Instead, Israel in every case emerged as the most beneficial party at the end by seemingly neutralizing their conflict with other Arab countries, gaining acceptance as a sovereign state in this region, and reinforcing its status quo borders and utilization of water resources, making it sound as if the only issue remaining is the Palestinian issue which can be resolved in separation from the overall Arab-Israeli conflict. The rather stagnant relationship that characterizes the current Palestinian-Israeli relationship is an amalgamation of a series of peace talks and a multitude of diplomatic effort that spanned throughout the past century and into the twenty first century. These diplomatic efforts can be organized into three major eras, including pre-1967 era, post-1967 era, and the modern era, as shown in Fig. 1.

1 The Pre-1967 Era: A Hotbed of Occupation, Conflict, and Discursive Hegemony

Negotiation attempts between involved parties in Palestine as a way to scuttle conflicts took multiple tracks and covered more than just the contentious political aspects. Issues related to natural resources allocation, as an extension to land, sovereignty, and national identity, were also part of these negotiation attempts. Water in particular was central to these negotiations. Because of its importance to both human survival as well as economic development, water can fuel particularly contentious political conflicts but if carefully managed can also induce sustainable cooperation (Abukhater 2013). However, the previous water negotiation as well as the overall diplomatic efforts were proven unfruitful and ineffective in managing the Arab-Israeli conflict over disputed scarce water resources or even resolving the Palestinian-Israel dispute. In the wake of the establishment of the state of Israel and the forced displacement of the indigenous Palestinian population from their ancestral homeland, a grinding military occupation was imposed on the majority of the Palestinians. During that era, the Palestinians were not viewed by either Israel or the United States as a party to negotiate with but rather as simply Arab inhabitants of a land that was largely empty, perpetuating the lie that Palestine was "a land without a people for a people without a land." This deliberate use of such terminological rules aimed at producing corrupted phraseology to cloak the real nature of the conflict and discredit any claims that the Palestinians may have to their own land. By creating this communicative and discursive hegemony, the Zionist movement was successful in placing the Palestinian narrative at a lasting disadvantage, assiduously providing Israel the discursive indemnity it needed once it was established in 1948 (Khalidi 2013). Moreover, the United States policies towards Israel and its geopolitical acuity it deployed to provide cover for its occupation of Palestine have all made the Israeli occupation politically "invisible." This was amalgamated with the supposed Israel's security which has always been at the forefront of any political discourse, and has been often used to justify Israel's actions towards the Palestinians, whether it was legitimate or not. As such, the Palestinian human rights, sovereignty and security were not even given any meaningful consideration. This structure furthermore made it possible to depict Israel as a resounding success story, while repeatedly attempting to cast doubts regarding the Palestinian narrative. Instead, their image has been repeatedly shattered and dispersed as a social and political entity. The reality that this structure has been trying to distort is the fact that the progressively exacerbated Palestinian tragedy has been a direct and unfortunate result of the supposed Israeli success (Khalidi 2013). Palestinians under Israeli occupation and the newly-formed Israeli domination of their land have been deprived from their very basic human rights. During the years following the establishment of Israel, the overall perception of the conflict has been tightly controlled and deliberately misconstrued, which all culminated in the perceived "intractable by design" state of the conflict that has been exacerbated by the lack of a U.S. diplomatic discipline to exert any substantive influence on Israel to abide by international laws. According to Rashid Khalidi, this carefully constructed realm of obscurity which was manifested in the deliberate misuse of language "has thoroughly corrupted both the political thought and action" (Khalidi 2013).

As this glaring reality was apparent to many, the diplomatic effort on the political side was hampered by this discursive dominance and hegemonic imbalance that was assiduously created to indemnify the actions of Israel in taking over the country and its resources. This power-structure imbalance was a key contributor to the apparent feebleness of these negotiation efforts and the resulting agreements that ensued. As the Palestinian literary and cultural critic Edward Said astutely described it in 1988: "It is by no means an exaggeration to say that the establishment of Israel as a state in 1948 occurred partly because of the Zionist acquired control of most of the territory of Palestine, but partly because they had already won the political battle for Palestine in the international world in which ideas, representation, rhetoric and images were at issue."

On the water front however, the long process of negotiation between the disputing parties and the several hydro-political and geopolitical events date back to the early 1900s. Sporadic diplomatic effort occurred to produce only bilateral agreements in order to manage the sharing of regional water resources, including both surface and groundwater. Throughout history, several water schemes were prepared to utilize the waters of the Jordan River; starting as early as 1899 with the Zionist Organization and later under the Mandate (Naff and Watson 1984). For example, some of these bilateral agreements to govern the sharing of the Jordan River water were reached as early as 1920, involving Britain and France. This agreement over the Jordan, Tigris-Euphrates, and Yarmouk waters governed the Franco-British convention on certain points connected with the mandates for Syria, Lebanon, Palestine and Mesopotamia. However, border disagreements between them during the demarcation process resulted in decades of bitter conflict over these international borders. In 1923 the two parties again signed an agreement about the boundary line between Syria and Palestine from the Mediterranean to El Hamme. Shortly after that, the 1926 agreement to facilitate good neighborly relations concluded between the British government on behalf of Palestine and the French government on behalf of Syria and Great Lebanon. In 1953, Jordan and Syria signed an agreement concerning the utilization of the Yarmuk waters. Although these attempts constituted crucial junctures that opened the door to prospects of enhanced future cooperation and diplomatic relations utilizing water as the basis on which to bring nations together, the issue of the Palestinians was not properly addressed. Even after the establishment of Israel and the consequent events that led to the displacement of the Palestinians and destruction of their villages and towns in 1948, this issue oddly remained unresolved.

In the post-1949 period, the Arab-Israeli negotiation process underwent many stages but due to the prevailing political atmosphere that was less than conducive these cooperative efforts to govern the sharing of the regional water resources did not occur openly. For example, Israel and Jordan, with the United States intervention, were engaged in a series of extensive secretive talks to manage the sharing of the Yarmouk water that date back to the 1940s to avoid criticism (Sosland 2007). This secretive nature of the diplomatic effort and the constructive confrontation continued to dominate the countries' political relationships for many years as the conflict was not deemed "ripe" for open negotiation (Burgess and Burgess 2007).

After the establishment of the State of Israel in 1948, there was a pervasive bias for unilateral action and little diplomatic relationship. This period which witnessed a stagnant diplomatic scene can be best characterized by unilateral actions on both sides of the isle but for different reasons. Both Israel and the neighboring Arab countries were actively and unilaterally creating national water development plans to satisfy their respective water needs and improve their economies. On one hand, Israel driven by its territorial appetite and the desire to attract an increasing number of Jewish immigration, was fixated on unilaterally transferring as much water into these areas witnessing increasing population growth, even if that meant the need to engage in armed conflicts. On the other hand, the surrounding Arab countries were preoccupied with the need to unilaterally find ways to foster their capacity to accommodate the increasing numbers of Palestinian refugees forced to flee their country (Lowi 1995; Shamir and Haddadin 2003). Ironically, only unilateral solutions were proposed and pursued to resolve a fundamentally regional problem. During these early years, there was little or no direct communication between the parties and a series of conflicting plans were unilaterally proposed to govern and manage the utilization of the Yarmouk River water. These unilateral proposals included the McDonald Plan and the Bunger Plan (both proposed by Jordan), the All Israel Plan, and the Syrian-Lebanese Plan (Haddadin 2002). A fundamental difference between these plans is concerned with their proposed in-basin use or inter-basin transfer of water (Haddadin 2002). Reflecting a Zionist vision that was originally developed by Theodore Hertzl and adopted by Bourcart in 1899, plans proposed by Israel envisioned constructing its National Water Carrier to divert water for out-of-basin uses in the coastal plain and the Negev (Shamir and Haddadin 2003). On the other hand, the McDonald Plan assertion that "the general principle, which to our mind has an undoubted moral and natural basis, is that the waters in a catchment area should not be diverted outside this area unless the requirements of all those who use, or genuinely intend to use the waters within the area have been satisfied" was the basis for the plans proposed by the Arab countries (McDonald 1951 cited in Haddadin 2002). Not only did both of these plans fail to address the overall regional and multilateral nature of the shared water resources in a comprehensive fashion in order to adequately satisfy the overall water needs of all parties, but they also introduced additional sources of tension, as they provided national solutions to a regional problem. This situation caused a number of armed clashes along the borders in the Armistice Demarcation Zone (ADZ) that was established by the Armistice Agreement in 1949 (Lowi 1995; Haddadin 2002).

Table 1 Water allocations of different development plans, in MCM/year

Plan	Israel	Jordan	Lebanon	Syria
Main	393	774	–	45
Cotton (Israel)[a]	1290	575	450	30
Arab	182	698	35	132
Unified	400[b]	720[c]	35	132

[a]Cotton Plan included integration of the Litani River into the Jordan Basin
[b]Unified Plan allocated Israel the "residue" flow, what remained after the arab states withdrew their allocations, estimated at an average of 409 MCM/year
[c]Two different summaries were distributed after the negotiations, with a difference of 15 MCM/year. On allocations between Israel and Jordan on the Yarmouk River. This difference was never resolved and was the focus of Yarmouk negotiations in the late 1980s
Data sources Adapted from Naff and Matson (1984), Wolf and Newton (2008) (Product of the Transboundary Freshwater Dispute Database, Department of Geosciences, Oregon State University. Additional information about TFDD can be found at: http://www.transboundarywaters.orst.edu)

As the parties resorted to diplomatic means in an attempt to cooperatively manage the development of the Jordan River Valley and settle the issue of the Palestinian refugees, a number of plans were proposed and considered. These plans include the Main Plan, the Cotton Plan, the Arab Plan, and the Unified (Johnston) Plan. Although the goal of these plans was to suggest different amounts of water allocation to each country involved, it was clear that the Palestinian water needs were utterly ignored. These volumetric water allocation schemes are listed in Table 1.

The Main Plan was carried out by Chas Main and the Tennessee Valley Authority (TVA) upon the request of the United Nations Relief and Works Agency (UNRWA) in 1953 constituted the basis of the later Shuttle Diplomacy. This plan excluded the Litani River from the system and instead proposed the construction of a series of small and medium size dams on the Hasbani, Dan, Banias, and Maqarin while utilizing Lake Tiberias for storage (Haddadin 2002). Israel, reacting to the Main Plan, proposed the Cotton Plan in 1954, which incorporated the Litani River to artificially augment the size of the supply and further concurred on the use of Lake Tiberias as the main storage facility. Unlike the Main Plan, the Cotton Plan was criticized on the basis that it required out-of-basin water transfers. As a counter proposal to the Main/TVA Plan, the Arab League Technical Committee in 1954 proposed the Arab Plan which excluded the Litani River, promoted in-basin use, and uniformly rejected storage in Lake Tiberias as it is entirely located in, and controlled by, Israel (Shamir and Haddadin 2003). Although were envisaged to provide a mechanism to manage the use of shared water resources, these separate and conflictive plans emerged as a source of additional tension in the region that eventually called for reconciliation. With tension on the rise, the disputing parties opted to resort to diplomacy as a way of moving beyond their disagreement on the mechanism of water allocation proposed by these plans.

The various reconciliation efforts that started in the 1950s resulted in a diplomatic process and a series of cooperative arrangements, albeit mostly bilateral. The United

States special envoy to the Middle East, Ambassador Eric Johnston, conducted the first multilateral cooperative venture in 1953–55, known as the Shuttle Diplomacy (Lowi 1995). As was noted above, the establishment of the State of Israel in 1948 prompted the two main co-riparians most dependent on the Jordan River system, Israel and Jordan, to undertake unilateral actions for the development of the river system. Not only did they have drastically different and contradictory visions of how the river ought to be developed, they also spurned the plans proposed by the other riparian heralding a heightened level of tension that could potentially threaten the overall stability of the region (Zawahri 2009). As a result, the U.S. ipso facto took a more proactive role in promoting a regional approach through a coordinated diplomatic effort. The involvement of the United States was motivated by many reasons and a number of events. These included protecting its interests and mitigating the communist influence in the region while assuring Israel's security. Moreover, as the Arab-Israeli conflict intensified, so did the Soviet-American rivalry, which resulted in an arms race in the Middle East. These profound geopolitical events had a defining impact on the regional stability and security, which in turn led to a notable shift in the U.S. diplomacy towards a more proactive role, known as the "piecemeal approach," in settling the Arab-Israeli dispute (Sosland 2007). Declaring this "piecemeal approach" adopted by the U.S. and highlighting the intended equitable nature of the Plan, the U.S. administration contends: "In the period 1953–1955, the United States Government, through the work of Ambassador Eric Johnston, made a strenuous and generous effort to devise a plan for unified development of the Jordan Valley that would be equitable and acceptable to all parties. Regrettably, the comprehensive [Johnston] Plan drawn up by Ambassador Johnston failed to win the immediate approval of the parties. However, in the absence of later studies of an equally comprehensive nature, we have considered that it remained useful as a model for full, orderly, and equitable utilization of an immensely important international water resource. We remain convinced of the benefits of the types of unified development embodied in the [Johnston] Plan. In the absence of agreement thereon, we have supported the right of the riparians to proceed with national water programs provided these did not conflict with the general principles and allocation patterns envisioned in the [Johnston] Plan" (Whiteman 1964).

Despite the importance of this mission to the U.S. which was fully cognizant of the serious ramifications and impending potential for a regional conflict because of the parties' aforementioned unilateral actions, the Palestinian interests and security concerns were not a motivating factor for this diplomatic intervention by the U.S., and as such these interests and concerns were not considered. The importance of these U.S. led efforts was expressed by President Eisenhower, who described the mission to be of "primary importance to the United States" (Slany et al. 1986, cited in Sosland 2007). Through his Shuttle Diplomacy, Ambassador Eric Johnston embarked on a regional reconciliation effort on October 7, 1953. He met separately with representatives of Israel and the Arab League Technical Committee (under Egyptian leadership) to manage the sharing and development of the Jordan River Valley. These separate meetings focused on each country's irrigation and agricultural development needs to avert potential conflict that could arise as a result of the lack of coordinated regional effort to manage the development of these shared water resources (Louka 2006).

This Shuttle Diplomacy aimed at developing an integrated plan for the regional development of the Jordan River system and to comprehensively settle the regional dispute through multilaterally allocating water resources in order to foster cooperation and economic stability. The objective of the mission, as stated by the Secretary of States John Foster Dulles, was "to secure agreement of the states of Lebanon, Syria, Jordan, and Israel to the diversion and use of the water of the Jordan River basin" (Slany et al. 1986, cited in Sosland 2007). The Johnston's mission, akin to the Main Plan, purposefully ignored the political boundaries and focused on the trans-boundary aspects of water to reach a resolution that can be accepted by all instead of facing political complications related to boundary disputes between Israel and the surrounding Arab countries. By so doing, it avoided political hurdles related to Israel's boundaries but by the same token it ignored the needs of the Palestinians who were also a fundamental party involved in the dispute.

Johnston's negotiations were based on key foundational elements, including the utilization of the TVA/Main Plan as the basis for water distribution, the adjustment of the Armistice Line to mitigate Israel's exclusive control over Lake Tiberias, the elimination of the demilitarized zones in the Jordan Valley, and the exclusion of the Litani River as an exclusively Lebanese river. Of particular consideration in the proposed water allocation scheme were several factors, including (1) the area of the arable land; (2) the assumed cropping pattern; (3) the cropping intensity; and (4) the water duty per unit cultivated area (Shamir and Haddadin 2003).

This cooperative effort was hampered by a number of challenges related to (1) the quantitative division (water quotas) allocated to the different co-riparians considering the very limited discharge of such small stream; (2) incorporating the Litani River into the Jordan system as a way to augment the size of the overall shared water system; (3) the intended Israeli use of the Jordan River water for areas located outside the basin; (4) the utilization of Lake Tiberias as a floodwater storage facility; and (5) the issue of international supervision. The relatively insignificant amounts of water volumes available in the river posed a significant challenge to reaching an acceptable unified plan able to allocate water quotas and further warranted finding innovative ways to augment the size of the resource in question. Israel advocated for the idea to augment the river system water supply by simply including the Litani River as part of the river system. Due to the fact that the Litani River originates and entirely flows within Lebanon, this idea was rejected by the Arab states who viewed it as an Israeli political maneuver to capture as much water as possible from the Jordan River system. After several rounds of shuttle diplomacy, the parties agreed to exclude the Litani River from the negotiations on the basis that it is an entirely Lebanese river and was treated as separate from the Jordan River system. Another idea that was considered to augment the water but was mutually spurned by both Arabs and Israelis for different reasons was to utilize Lake Tiberias as the main storage reservoir for the Yarmouk floodwater accompanied by dam construction along the River. For Israel, utilizing Lake Tiberias as a storage facility meant compromising its sovereignty of the Lake. For the Arab states, this constitutes a threat to their water security because it meant an increasing dependency on the Israeli controlled Lake Tiberias to obtain water for their own use. However, both sides acquiesced to the dam construction proposal and utilizing the Lake for an aggressively reduced amount of floodwater storage.

The intended use of water by the parties emerged as a major point of contention. The Arab States intended to utilize the water for intra-basin water use, whereas Israel intended to use the water outside the basin, namely in the coastal plain and Negev Desert, which in turn meant an inter-basin water transfer plan. Although the intended Israeli water use was criticized and rejected by the Arab States which viewed it as out-of-basin water transfer, they eventually agreed to it. The negotiations were also concerned with establishing an institution to govern the water management of the river basin and monitor implementation and compliance. Despite the significant disagreement among the parties and the politically controversial and consequential issues on the table, the shuttle diplomacy efforts of Johnston were able to develop a formula, known as the Unified Plan, for water allocation among the different co-riparians (as shown in Table 1). However, accepting the Plan by the Arab League Council meant ipso facto recognizing Israel. As such, although the technical committees from both sides accepted the Unified Plan, the Arab League Council rejected it in October 1955. Israel notwithstanding the parties' overwhelming opposition, resumed its diversion plans of the Jordan waters from the northwest corner of Lake Tiberias, via its National Water Carrier, which became fully operational in 1964.

It is crucial to note that the failure of the Johnston mission was due in large part to the challenges imposed by complex geopolitical realities that surrounded the issue and the timing of the diplomatic effort that was less than conducive. As the Plan was never ratified by any of the parties, it was never considered a binding agreement and instead remained simply as a "plan." The failure to ratify the Johnston Plan was related to a fundamentally systematic flawed in the engagement with the Palestinians that were clearly excluded from these negotiations. As a result of this root-cause, a number of issues were at play, including (1) the timing and macro-political nature of the conflict, namely the Arab states' view that recognizing the right of Israel to exist at that time may compromise the Palestinian cause and right of return, and (2) the lack of economic incentive for the Arab states to adopt the agreement, due to its high cost when compared to its immaterial benefit. This was particularly true for Lebanon which could potentially gain only 35 MCM by adopting the Plan despite the fact that it would have been a major economic undertaking (Zawahri 2009).

Although the Plan was never ratified, it became a customary law in the region and was commonly accepted by all Arab states. The Plan constituted the basis for the Jordan River Valley development effort, as both the Arabs and Israelis continued to informally adhere to its technical details and water allocation schemes, until the 1967 War, in which Israel expanded its boundaries, and as a result, its water consumption and utilization. The Plan is still referred to as the foundations for any future negotiations and the basis for any potential water-related allocation along the Jordan River Valley, although it was criticized for its exclusion of the Palestinians and groundwater resources. The Plan was the catalyst for many reconciliation attempts, namely the Picnic Table Talks between Israel and Jordan, the "water-for-peace" desalination projects during the 1960s, and the negotiations over the Yarmouk River and the Unity

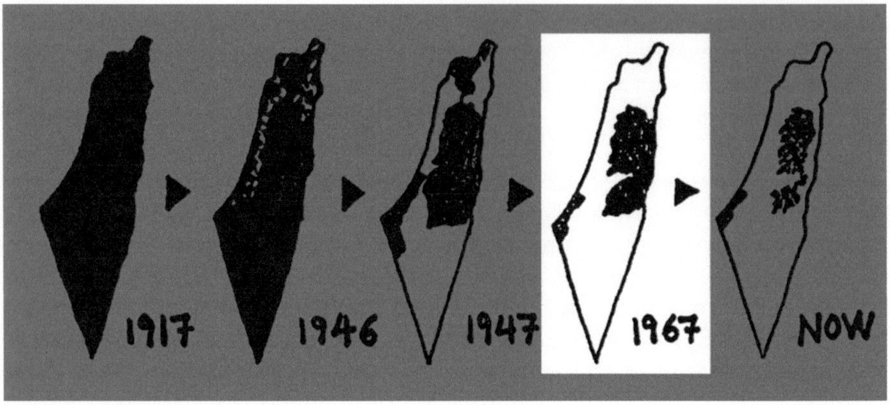

Fig. 2 A map of Palestine in 1967

Dam that took place during the 1970s and 1980s. For Israel and Jordan, the Plan had a particular importance serving as the guidepost for the development of the River basin. Because Israel and Jordan were the greatest beneficiaries of the Plan as well as the most dependent co-riparians on the Jordan water, they made their intention to unilaterally comply with it, clear to the U.S. through secret notes (Sosland 2007). The commitment of Israel and Jordan to abide by the Plan was in exchange of U.S. economic assistance in the development of the National Water Carrier in Israel and the East Ghor Canal in Jordan (the King Abdallah Canal). This was evident as the U.S. aid was conditional and closely tied to the extent to which the two parties abide by the amounts specified by the Plan, indicating in a letter sent to Jordan that, "assistance will be extended provided there is an explicit undertaking from the Hashemite Kingdom of Jordan that it will not draw from the Yarmouk River more than the share allotted to it under the [Johnston] Plan" (cited in Sosland 2007).

Notwithstanding its shortcomings, the Plan was successful in significantly ameliorating the hydro-political tension over the waters of the Jordan at that turbulent time. Although the U.S. involvement in this era was timid, its diplomatic effort was confined to protecting its interests in the Middles East coupled with its consistent and firm support of Israel in every decision it makes or action it takes. Although the U.S. rhetorically supported the Palestinian refugee right of return and compensation as represented in the 194 UN Resolution, that support was never taken to the level of implementation and further ended in 1992. These rights were completely ignored in the UN Security Council Resolution 242 of November 22, 1967 which has been used as the foundational principle for the peace negotiations for many decades. The Palestinians were not even mentioned by name in this resolution which presumably aimed at a "just settlement of the refugee problem" and is regarded as the basis for the peace process. These diplomatic efforts that took place during this era failed to either change the map of Palestine (as shown in Fig. 2), or resolve the issue of water allocation for the Palestinian communities.

2 The Post-1967 Era: Political Obfuscation and Diplomatic Feebleness

The 1967 Six-Day-War was yet another decisive victory for Israel which was able to expand its territories even further with the help of the unwavering military and political support of the U.S. In the wake of this war, Israel was able to take over the remainder of historic Palestine, including the West Bank and Gaza, as well as the Egyptian Sinai Peninsula and the Syrian Golan Heights. As such, Israel's territorial objectives were made manifestly clear to the rest of the Arab countries involved in the conflict. The United States policies towards the Middle East, namely its unequivocal support of Israel and sporadic, timid, and qualified expression of concern about the Palestinian rights during the Carter admiration, did not change much. In some very limited occasions, it even indorsed the Palestinian rights to self-determination but was never actively involved in translating this support into concrete actions on the ground. For many decades, the U.S. did not even wish to recognize or conduct any direct negotiations with the PLO, despite the fact that it was recognized by the UN as the "sole and legitimate representation of the Palestinian People." On the other side, the U.S. was actively involved with Israel and in 1975 was able to negotiate a Memorandum of Understanding in which it committed not to "recognize or negotiate" with the PLO unless it recognizes Israel's right to exist (without remanding the same reciprocation from Israel), renounce terrorism, and accept the UN Security Council Resolution 242 and 338 (Khalidi 2013). The U.S. maintained this position towards the PLO until the latter met these conditions in 1988. President Jimmy Carter was believed to be relatively the most balanced administration that spoke of "self-determination" for the Palestinians. However positive this narrative might have been, the issue of Palestinian self-determination was never fully addressed and went quietly into utter occultation as the Cold War intensified.

One of the most remarkable diplomatic moments in this era of political uncertainty is the US-brokered 1978 Camp David Accords commandeered by the Egyptian President Anwar Sadat and signed by President Jimmy Carter and Israeli Prime Minister Menachem Begin in September 1978. The Accords proposed a Framework for Peace in the Middle East, which was condemned by the United Nations General Assembly as it once again failed to engage the Palestinians, and for a Peace Treaty between Egypt and Israel which amounted to the 1979 Treaty. Although the 1978 Camp David Accords were deemed one of the most serious American diplomatic effort to address the issue of Palestinian autonomy, these provisions were not translated into the 1979 Treaty or the ensuing negotiation effort that took place between 1978 and 1982.

In an unprecedented event of its kind, the Egyptian President Anwar Sadat boldly gave a speech at the Israeli Knesset in November, 1977 where he expressed his desire, and outlined a plan, for peace with Israel, after notifying the U.S. of his intention to seek a "peace agreement," setting in motion a series of events that led to the 1978 Camp David Accords and culminated in the 1979 Peace Treaty. Growing progressively impatient with the dilatory diplomatic approach of the Carter administration, President Sadat was only interested in a bilateral resolution in order to resolve their conflict and put an end to the state of hostility between his country and Israel, independent from the overall Arab-Israeli conflict. His primary objectives of his unctuous cooperation with the U.S. and bilateral talks with Israel were to regain Egyptian sovereignty over the Sinai Peninsula with international guarantees, end the state of hostility with Israel, and maintain cordial diplomatic relationship with the U.S. This move demarcated a clear departure from the position taken by his predecessor, Gamal 'Abdel Nasser, who adopted the "land for peace" Security Council Resolution 242 (Khalidi 2013). Due in large part to the influence of the Cold War and the impact of the 1973 Six-Hour-War in which the Egyptian military took over the Suez Canal, the U.S. opted to engage with Egypt and reluctant Israel in peace talks. These peace talks, which were driven by the U.S. desire to maintain its influence in the region and block the Soviets from doing the same, were obviously not interested in advocating for inalienable Palestinian national rights. These negotiations, although perceived by Israel as concessions, eventually gave Israel a great regional and international advantage, by isolating and relegating Egypt's desiderata and securing an increasing U.S. financial support it receives respectively. Once again, Israel emerged as the most advantaged party in the wake of this treaty.

Although the Carter Administration advocated for the Palestinian right and even imposed its well to include "the legitimate right of the Palestinian people and their just settlement" in the agreement, it eventually capitulated to Israel's will and meekly accepted its terms regarding the Palestinians. These were very restrictive terms that permitted the Palestinians of the Gaza Strip and the West Bank a five-year autonomy period under full Israeli control, excluding the larger Palestinian refugees in diaspora, Jerusalem and inside Israel (Khalidi 2013). Reflecting a legacy of continued systematic exclusion of the term "Palestinians" that dates back to the Balfour Declaration of 1917, the Carter Administration later explained that the term "Palestinians" was construed as "Palestinian Arabs." The agreement did not accomplish what Carter intended, which was an end of Israeli occupation of the Palestinian land, Palestinian autonomy, and a halt of Israeli settlement expansion none of which was actually realized under this agreement. Instead, this agreement helped in acknowledging the legitimacy of Israel and its occupation of the Palestinian lands. Worse, the establishment of this "Self-Governing Autonomy" for the Arabs of the Gaza Strip and the West Bank was never implemented, personifying a reprise of lack of interest in the Palestinian issue throughout most previous diplomatic efforts. As a result of this bilateral treaty between Egypt and Israel, the latter started to withdraw its troops from the Saini Peninsula which was fully returned to Egypt in the spring of 1982.

It is crucial to note that although the 1979 Peace Treaty made direct recommendations for the Palestinians, the PLO was never represented in any of these talks leading to the agreement for a number of reasons (Fig. 3). Each party involved in these talks had its own distinct reasons for the systematic exclusion of the Palestinians from these negotiations, which ultimately constituted a "Gordian knot" for its implantation and overall acceptance and was further the basis for its failure (Fig. 4).

First, the United States considered the PLO a terrorist organization and further made a commitment to Israel not to involve it in any negotiation—a position that the U.S. held until the 1993 peace process started. This tactic of labeling Palestinian

Fig. 3 Cartoon by Le Pelley, "waiting for admission to the Peace Conference on the Middle East". *Source* The Christian Science Monitor (1977)

organizations involved in acts of violence against Israel (regardless of whether they illegally target civilians or legally defend themselves against illegal occupation) is also used today to describe Hamas for its role in fighting the Israeli occupation. The same narrative is not equally used to describe the documented war crimes that Israel

Fig. 4 Cartoon, "agreed not to use the sword". *Source* Library of the Congress (2018)

has been committing throughout its history since its establishment to modern day, which reveals the systematic and deep-rooted U.S. bias (Khalidi 2013). Most notable of these atrocious acts that Israel committed against the Palestinian communities were around the time of its establishment which resulted in the displacement of many Palestinian towns and villages and deprived its local inhabitants languishing in refugee camps the opportunity to ever go back to them. Most recent of these practices were the widely documented war crimes Israel committed in Gaza in 2008/2009, 2012, and again in 2014, which were all overlooked by the U.S and were depicted in its public discourse as simply Israel "defending itself." Second, for Israel, the term "Palestinians" was in question as it denied its very existence. As such, Israel was not willing to involve the Palestinians in any talks that may legitimize them or their cause. Nothing can illustrate this Israeli position towards the Palestinians better than Golda Meir, fourth Prime Minister of Israel, who in 1969 enunciated:

> There were no such thing as Palestinians. When was there an independent Palestinian people with a Palestinian state? It was either southern Syria before the First World War, and then it was a Palestine including Jordan. It was not as though there was a Palestinian people in Palestine considering itself as a Palestinian people and we came and threw them out and took their country away from them. They did not exist.[1] (Gilad 2015)

This, akin to many Israeli leaders' assertion of their views of the Palestinians, reveals beyond a shadow of a doubt their rejection of the very existence of the

[1]Quoted in Sunday Times (June 15, 1969), also in The Washington Post (June 16, 1969).

Palestinian people and by extension their utter denial of any Palestinian claim to historic Palestine. One of the most revealing aspect about the way that Israel views the Palestinians is their continual referral to the West Bank as "Judea and Samaria" which is indicative of their disdainful denial of the Palestinian claim to the land of Palestine. To them, the land of Israel, or "Eretz Israel" represents the entire area from the sea to the Jordan River that was under the British Mandate.

Third, the Egyptian President Anwar Sadat was interested in simply resolving its individual dispute with Israel in separation from the larger political and territorial conflict with the rest of the Arab countries. He saw the Israeli-Palestinian conflict as a much more nuanced and complicated dispute for his country to represent or tackle. He was concerned that it may hinder his country's ability to reach any resolution if the talks were also concerned with the issue of Palestine. As such, not interested in diluting the focus of the negotiation, he narrowly sought to bilaterally settle the dispute between Egypt and Israel and regain sovereignty and control over all Egyptian land including the Sinai Peninsula that was occupied by Israel in 1967. For Sadat, he wanted to be viewed as a victorious Arab leader after the 1973 War, Israel needed security and legitimacy, while the U.S. was interested in fostering its influence and protecting its strategic interests in the Middle East from the Soviet influence. This combination was proven potent enough to incentivize the parties to engage in peace talks and further create an atmosphere conducive to produce an agreement they can all endorse. However, none of the major issues surrounding the Israeli occupation of the Palestinian land, continued settlement expansion, or Israel's notorious violation of international law were discussed or addressed.

Although it was successful is accomplishing the aforementioned goals, the treaty which had a significant impact on the whole region failed on many fronts. The most significant aspect of this treaty, however, is not simply its direct substantive impact on both of the disputing parties, but rather the far-reaching perception that it created. It ushered in a new era of peace and normalization with Israel, which at the time of the treaty was not acceptable or even fathomable by any of the parties involved in the conflict. By creating a diplomatic hype, the U.S. in particular was quick to celebrate it as an unprecedented diplomatic achievement and a significant step towards resolving the formidable Middle East dispute. The reality, however, was quite the opposite (Fig. 5).

While it ended the state of war between Egypt and Israel, the 1979 Peace Treaty had detrimental effect on the regional stability causing more violence in both Palestine and Lebanon which continues to the present day. It excluded the Palestinians from the negotiation, failed to implement any of its stipulations about the Palestinian rights to self-determination and autonomy, perpetuated the issue of the Palestinian refugees, and contributed to far greater violence in the region, to name a few. Although the implementation of this Peace Treaty has been held by both parties despite considerable upheaval in the region, the relationship between the two countries has not been particularly warm as a result of signing the treaty. The most formidable issue was not the signing of the treaty itself but the failure of the leaders (namely President Sadat) to gain acceptance and support from their own people. In fact, many link the assassination of the Egyptian President Anwar Sadat to his controversial role in this

Fig. 5 Cartoon, "meeting his enemies has been safe enough—it's going back to his friends that takes guts". *Source* The Daily Mail (1977)

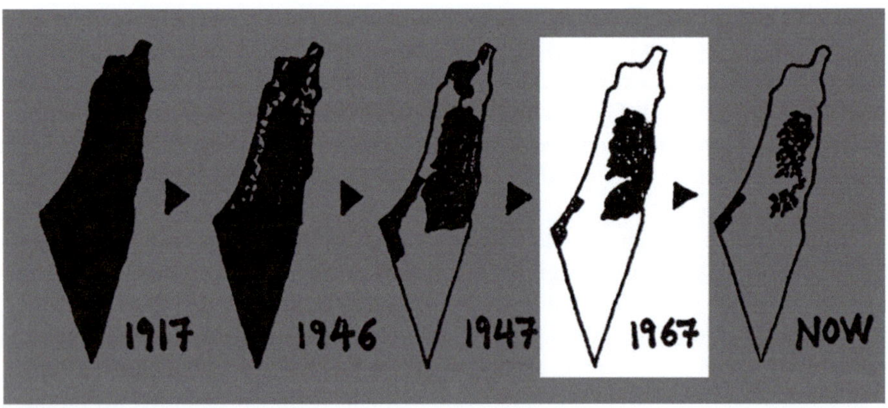

Fig. 6 A map of Palestine in 1967

agreement, which was viewed by most Arabs and Palestinians as a sellout and a betrayal of the inalienable Palestinian rights to have a state of their own (Fig. 6).

The American diplomatic effort led by the U.S. President Regan (later known as the Regan Plan) to react to the Israeli invasion of Lebanon and the massacre of nearly fifty thousand civilians resulted in a cease-fire on August 12th, 1982 and a series of events that were culminated in the evacuation of the PLO leadership from the Lebanese capital. The Regan Plan, which opposed the occupation of the West Bank and Gaza Strip, was rejected by both Palestinians and Israelis. Israel was infuriated as it called for the "immediate adoption of settlement freeze" regarding them as unnecessary to the security of Israel and would ultimately impede any fairly negotiated outcomes. Particularly revealing in this context is the polarization of the Arab world between Arab nationalist and pro-American governments in the wake of the Cold War. Despite its proclaimed opposition to the occupation of the West Bank and Gaza Strip, the U.S. viewed Israel as a strategic asset in the region in light of the Cold War and was not interested in exercising any pressure to force Israel to withdraw from the territories it recently occupied (Khalidi 2013). As such, the end of this era and the result of the diplomatic effort did not impact the map of Palestine that effectively remained under Israel's occupation since 1967.

Not much diplomatic effort took place since the 1979 Peace Treaty which was proven to be a diplomatic debacle as its failure in resolving the core issue of Palestine became apparent with every passing year. During this period, the conflict has been particularly violent despite the many intermittent attempts to cement a lasting resolution. The Palestinians continued to live under a grinding military occupation in the Gaza Strip and the West Bank as refugees with no basic rights. As a result of the mounting pressure and human rights violations against the Palestinian people, a remarkably transformational event took place in 1987 known as the first Palestinian uprising (Intifada). The Palestinian Intifada irrupted as a declaration of the Palestinian people's rejection of Israel's brutal occupation of their territories. It introduced an element of large-scale popular resistance that did not exist in years past.

In 1988, the PLO Chairman, Yasir Arafat declared the establishment of the state of Palestine, renounced terrorism, and called for an international peace conference involving Palestinians, Arabs and Israelis under U.N. auspices and based on the UN Resolution No. 181 and the Security Council Resolutions 242 and 338. With this declaration, Arafat took the decision to enter into a peaceful settlement with Israel, waging a new era of diplomatic effort that would take place in the early 1990s and would ultimately lead to the establishment of the Palestinian National Authority in the Occupied Palestinian Territories, as will be discussed next. At this point in history, the map of Palestine has not changed much despite the timid and intermittent effort to find a negotiated settlement that could sustainably put an end to the conflict (Fig. 7).

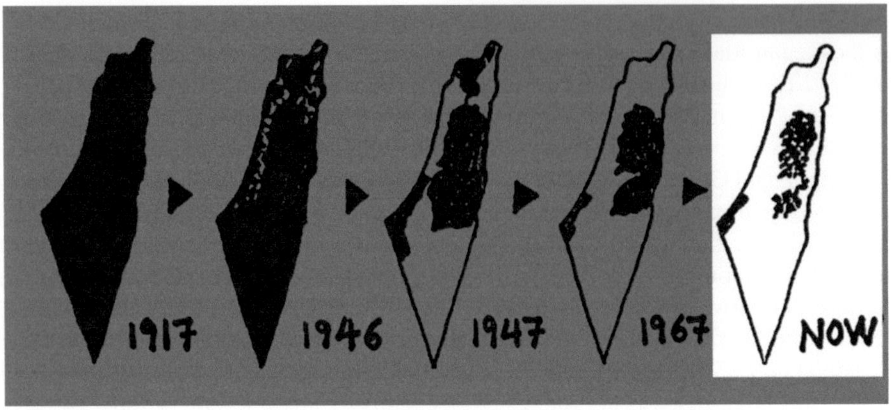

Fig. 7 A map of Palestine today

3 The Modern Era and the "Peace Process": From Clandestine Talks to Political Chaos

It is believed that the current Middle East peace process to settle the Palestinian Israeli conflict officially started on the White House lawn on September 13th, 1993, with the handshake of the Israeli Prime Minister, Mr. Yitzhak Rabin, and the Palestinian President Mr. Yasir Arafat. This handshake symbolized a renewed U.S. diplomatic interested and proactive role to demarcate a more serious turn of events regarding achieving a just, lasting and compressive peace between Israel and Palestine. Before the beginning of the official bilateral negotiation process, various attempts were made to reach a negotiated settlement, none of which yielded a discrete resolution. This situation changed by the 1980s with significant geopolitical events that made the prospect of a negotiated settlement plausible. There were a number of global events that marked a harbinger of a new era and paved the way to this moment, including the demise of the Soviet Union, the unipolar American dominance, the 1991 Gulf War to expel Iraqi forces from Kuwait, and the isolation of the PLO due to its ill-advised alignment with the Iraqi regime. These events provided impetus for the U.S. to resurrect the diplomatic effort to resolve the Arab-Israeli conflict. Motivated by its enhanced post-Cold war status and global preeminence, the United States spoke of a "new world order," and engaged in arduous shuttle diplomacy led by the U.S. Secretary of State James Baker who attempted to reinforce the U.S. global dominance through comprehensively settling the Middle East conflict. Part of these diplomatic efforts he conducted were eight trips during the eight months following the Gulf War that were the foundation for the Madrid Peace Process (Abukhater 2013).

In October, 1991, the PLO and the Arab states entered into direct peace negotiations with Israel as part of the peace conference in Madrid, where the U.S. saw an opportunity to accomplish its goals. In addition to its sponsors, the United States and the Soviet Union, the Conference also invited international presence, namely representatives from the European Union and the United Nations. The Madrid Conference agenda included items related to security, water, refugees and displaced persons, borders and territorial matters (Lillian Goldman Law Library 2008). Although the Madrid Conference brought all involved parties, including Israel, Lebanon, Syria, Jordan and Palestine, the Palestinians were not adequately represented reflecting the long-standing Israel's policy towards acknowledging or dealing with them. In addition, other neighboring countries also attended namely Egypt, Saudi Arabia, and Morocco, representing the Arab Cooperation Council, the Gulf Cooperation Council, and the Maghareb Union respectively. Under the auspices of the U.S., the Palestinians for the first time were allowed to attend the peace talks but were not represented as a separate delegation, reflecting a long-standing legacy of isolation and disinterest. Although most of the thorny issues on the table were concerned in one way or another with the Palestinians, they were represented as part of the Jordanian delegation at the beginning, but as the negotiation progressed they got separated into two delegations. Given the collapse of the Soviet Union in December of 1991, one of the cosponsors of the conference lost its influence and the whole conference was practically under full American control, which in turn was advantageous to Israel's interests and desiderata and effectively impaired the ability of the Palestinian negotiators to lead a meaningful start to the negotiations.

The negotiation between Israel and its neighbors ensued two tracks; (1) a multinational conference involving international participation that focused on issues of regional importance; and (2) separate, yet simultaneous, bilateral negotiation tracks between Israel and each of its neighboring countries to address common issues of concern (Fischhendler 2008). This deliberate Israeli-engineered segregation of the bilateral and multilateral tracks was a direct result of Israel's sulfurous position towards the Palestinian representation which was deeply abhorred within the Israeli political and public circles. Due to the fact that it was a key issue for the parties involved in these negotiations, water was a topic of negotiation in both the bilateral tracks and part of the multilateral negotiations. There was a clear separation between the water, a "low politics" issue, and the other negotiation over "high politics" issues such as security, mutual recognition, land, and borders, which evidently contributed to the failure of these efforts.

Five multilateral groups (representing issues of regional importance) emerged from the Moscow Conference, including water, environment, refugees, regional security and arms control, and regional economic development. The main goal of these multilateral talks was to meet the "other" party and facilitate common understanding and agreement regarding major issues, interests, needs, and concerns. On the water front, the agenda of the Multilateral Working Group on Water Resources had four important elements including (1) the enhancement of data availability, (2) water management and conservation, (3) enhancement of water supply, and (4) concepts of regional cooperation and management. Although the two bilateral and multilateral tracks of negotiation were in theory meant to complement one another, clashes occurred which hindered the efficiency of these negotiations and caused a great deal of mistrust from the very get go. These multilateral groups continued to conduct several rounds of negotiation in different locations, including Madrid, Moscow, Vienna, Washington, Oslo, Geneva, Beijing, Taba, Muscat, Athens, Amman and other sites (Shamir and Haddadin 2003). The goal of these talks as outlined by the U.S. "Letter of Invitation" was to formulate an "interim self-governing arrangement" for the Palestinians. This "interim" arrangement was extended and ended up forming the permanent state for the Palestinians to this day. According to Khalidi, 2013, "such prescriptions proceeded from the assumption that the problem was simply 'mistrust and suspicion' and lack of 'confidence' between two implicitly equal sides, rather than that one was in illegal military occupation of the territory of the other, and in effect had its boot on the other's neck." In reality, this transitional period served as a cover for Israel to consecrate and legitimize its occupation, entrench its settlement expansion, and worsen the conditions for the Palestinians living in these territories it occupies, proving that the very justification for this "interim" arrangement was patently false and equally abortive.

Critical in this context is the impact of these ground rules imposed by the U.S. on the Palestinian leadership, which was significantly detrimental to the trust-building process originally intended for this transitional period. Key issues such as ending the military occupation, halting and dismantling settlements, control of land and water, sovereignty, statehood, the status of Jerusalem and the issue of the Palestinian refugees were not allowed or addressed in these negotiations and were deferred to the final "permanent status talks." The Palestinians were not allowed to quibble over or even discuss any of these issues in a meaningful fashion as they were completely off the table and labeled as unwarranted "pre-conditions." Although verbally championed the halt of the Israeli settlements, the U.S. inability or unwillingness to translate this rhetoric into concrete actions to curb the ever-expanding Israeli settlements inside the Palestinian territories while negotiations were taking place was evident to the ill-advised and unprepared Palestinian negotiator. To that end, a proposal made by the head of the Palestinian delegation in Washington talks, Dr. Haydar 'Abd al-Shafi, to the PLO leadership recommending to disengage from the negotiations to respond to the U.S. duplicity regarding the Israeli settlements, was met with a huge political miscalculation on the part of the PLO. Despite the apparent signs and the clear warning from Dr. Haydar 'Abd al-Shafi, the PLO decided to continue the negotiations notwithstanding the fact that the U.S. studiously ignored its promises made in the "Letter of Invitation" (Khalidi 2013). His recommendation constituted a crucial moment of relative clarity in the fog of obfuscation that surrounded the Palestinian-Israeli relationship and the overall peace process. However, the PLO was desperate for any outcome that could potentially lead to any settlement or recognition, even if that meant accepting a deal on the basis of take it or leave it grounds. As such, they cast aside any warning signs and the obvious writing on the wall about the value of these negotiations. This position put the PLO and their negotiation team at a greater disadvantage as the weak, vulnerable and disparate party. Not surprisingly, this US-mediated and choreographed effort between 1991 and 1993 did not yield any agreements.

The parties were not able to make real progress during direct talks except through the channels of various secret negotiations that were commenced in December 1992 leading to the Oslo Accords or "Gaza-Jericho First" and the establishment of the Palestinian National Authority (PNA) (Klieman 2000). As a result of these clandestine talks, the PLO leadership in exile returned to the Gaza Strip and the West Bank in September of 1994 (Tamimi 2005). In these secret negotiations, Israel and the PLO affirmed mutual recognition, which resulted in the signing of the Declaration of Principles (DOP). One of the most critical flaws of this effort was the inability of the Palestinian negotiators to leverage and capitalize on the previous work of their predecessors. Instead, this work was completely overlooked and ignored in order to achieve quick and insignificant gains in the very short term without the ability to formulate a unified vision and strategic position to enable them to establish a meaningful progress on the key issues they have long advocated for. As the Israelis cling to the original Begin's agenda outlined in 1977, so did the PLO to the prospect of any negotiation notwithstanding the impending and obvious diplomatic miscalculations and any future ramification it may eventually cost them. It is worth mentioning that water issues were only discussed in the bilateral Israeli-Palestinian track of negotiation, but were never fully addressed or meaningfully resolved. The first paragraph of Article 40 "Water and Sewage" of the 1995 Oslo II agreement indicates that "… the Palestinian water rights in the West Bank … will be negotiated in the permanent status negotiations and settled in the Permanent Status Agreement relating to the various water resources" (The Israel-Palestinian Interim Agreement 1995).

In 1995, Israel and the PLO futilely attempted to negotiate *inter alia* the issues of borders, Jerusalem and settlements. Their intention however was to avoid causing changes to the situation on the ground without the final outcome of the permanent status negotiations. The Camp David Summit in July, 2000 was the first attempt at providing a negotiation platform over the "final status" issues. In April 2003, a peace plan for a two-state solution, known as the "Road Map," was devised by the Quartet (US, UN, Russian Federation and the EU) (Reinhart 2006). As it only offered the Palestinians "sovereignty without control," this grotesquely named "peace process" proved to be gravely deficient in making any substantive progress in the lives of the ordinary Palestinians who were eager to see real changes on the ground as a result of this process. At best, the supposed façade autonomy that the Palestinian were given in these confined cantons under the auspices of the two-state solution can be equated to the autonomy over how to decorate their prison cell.

As these hopes were proven to be utterly ephemeral, the Palestinian society saw the writing on the wall, literally on the separation wall and figuratively with the increasing strangulation of their existence. As a result, these negotiations were interrupted by the second Palestinian Intifada. The ever-subsiding mirage of real Palestinian sovereignty and statehood hampered the prospects of achieving any real outcome from these negotiation attempts. In addition, the U.S. inability or unwillingness to curb the unfettered expansion of Israeli settlements, coupled with the failure of the Arab countries to exercise any pressure on Israel towards that end, made it difficult for the Palestinians to take these negotiations seriously. While the parties were supposedly engaged in talks to reach an acceptable negotiated settlement, one side (Israel) was actively changing the contour of the land with earth-moving equipment to irrevocably impose a new reality that cannot be trivialized or ignored. This unabated settlement expansion, the isolation of the West Bank and East Jerusalem, and the erection of the infamous Israeli segregation wall inside the Palestinian Territories were the obvious visible realities that the Palestinians witnessed and had to contend with as a result of these negotiations. The devastating outcome of these negotiations can be clearly seen in the current state of affairs and the ever-shrinking map of Palestine, shown in Fig. 7. Not surprisingly, this evident failure of these negotiations to yield any value for the Palestinians was ultimately the impetus for the emergence of Hamas as a serious alternative and political contender to the PLO which is widely perceived as an extension to a failed process that was based on fundamentally flawed legal and political miscalculations.

In addition, these de facto boundaries of what constituted the potential Palestinian state in the "two-state solution" do not reflect the realty on the ground. The West Bank is sliced by the Israeli Wall, riddled with a network of illegal Israeli settlements that dominate the landscape, and further segregated into smaller areas that lack real connectivity, with severe restrictions on movement of people and goods. The Gaza Strip is under siege and completely cut off from the rest of the world. The current situation in the Occupied Palestinian Territories today does not allow the establishment of a viable Palestinian state, and anyone who promotes this two-state solution as a viable option might seem oblivious or unaware of the current situation on the ground and thus cannot be taken seriously. Unless serious consideration is given to dismantling the illegal Israeli settlements in the West Bank and East Jerusalem, ending the illegal siege on Gaza, and finding ways to connect the two Palestinian territories, real peace will not be insight anytime soon.

It can be said, however, that any future negotiation process that tends to completely discard or dismiss the two-state solution emerges as a premature and dangerous approach considering the tremendous support that it receives from many people on both sides of the divide and including the U.S. Whether two-state or one-state solution, it is not practical or fair to relinquish one option for the other, namely that neither is actually proven to be viable at this point. Instead, the two options must be considered *pari passu* in any serious future negotiations, with a comprehensive assessment of their viability, practicality, and acceptability.

The gradual abandonment of the peace process created a vacuum on both sides and a realization that peace is unattainable. On one hand, the radicalization of the Israeli society played a pivotal role in the exacerbation of this situation. On the other hand, the Palestinian society became accustomed to the peace process lofty promises without seeing real results, and as such, lost faith in the process and its brokers altogether. Any discussion of peace talks with Israel is viewed by them as a waste of time and a betrayal of their national struggle for freedom. This complete lack of trust in one another and in the peace process itself created a dangerous situation that needs to be addressed prior to any future negotiations.

From the Palestinian perspective, Oslo's legacy remained to be seen as a litany of unfulfilled promises. Almost approaching three decades to the ratification of the agreement, the situation on the ground is much worse with more than ever Israeli settlements, much less freedom of movement, and significantly worse socio-economic conditions for the Palestinians. The situation eventually escalated into three major wars on the besieged population of the Gaza Strip that resulted in the killing and displacement of thousands, and left the Strip in a situation that can only be described as disastrous and unsustainable.

On December 6th, 2017, demarcating the hundred year anniversary of the Balfour Declaration of 1917, U.S. President Donald Trump announced in an unprecedented move that the U.S. is officially recognizing Jerusalem as the capital of the state of Israel and the "Jewish people," reversing nearly seven decades of American foreign policy on the issue. He additionally ordered the move of the American Embassy from its current location in Tel Aviv to the fiercely contested city, defying all warnings from both allies and foes of the impending danger that this decision will bring to the region. This decision is problematic on many counts.

First, akin to the Balfour Declaration, it again constitutes a case of a promise made by those who do not own to those who do not deserve. As a result, this decision and the ensuing ceremonial move of the U.S. embassy to Jerusalem completely disempowered the Palestinian President Mahmoud Abbas who has been advocating for a peace settlement with Israel, making it impossible for him to justify to his own people any future peace negotiations with Israel when Jerusalem (the very heart of the conflict) has been "given" to one side. This move, which also marks the 70th anniversary of the *Nakba*, echoed a similar tone to that of the Balfour Declaration of 1917, whereby one country gives a second country the land of a third country. This is particularly true considering the fact that U.S. has no legal or ethical right to promise the city of Jerusalem to anyone else other than the Palestinians who own most of the city, let alone to Israel who occupied it by military force, and is still in violation of international law in this regard. Moreover, the U.S. administration, touting about it as a promise made, promise kept, may have compromised the peace process in the process. The political cost of this unilateral and very controversial move outweighs any short-term gains that President Trump may have gained with his hardline base. As this decision flies in the face of international law, U.S. allies who strongly opposed it, and the very basic principles of the peace process, it is neither in the interest of the Israelis nor the Palestinians, constituting a major blow to the prospects of peace. This is particularly true as Jerusalem is one of the key issues to be determined *via* direct negotiations between Israel and Palestine and cannot be dictated by anyone else. In this regard, author and religious studies scholar Reza Aslan astutely points that out in an interview affirming that "You can't have a negotiation over how to share a pizza, when one side is eating the pizza," equating Trump's move to "throwing a bomb into the mix" of an already inflammable conflict (Aslan 2017). With this in mind, it is difficult to imagine a positive outcome when it comes to a future peace settlement that recognizes the rights and obligations of both sides.

Second, declaring occupied Jerusalem the capital of Israel is not only illegal, but it will also undermine the prospect of peace in the region, and ultimately the Palestinian right to self-determination, sovereignty, and freedom. This decision has serious consequential implications on achieving just peace between the two sides. For many people, this decision means closing the final chapter on the death of the peace process, and practically puts an end to the two-state solution signaling a push towards a potential one-state solution. Contrary to Trump's groundless claim that this is a good move towards achieving peace by "taking Jerusalem off the table," this political maneuver is viewed by many as extremely detrimental to the overall peace process and the prospect of reaching any negotiated settlement between the two nations. Commenting on this, Saeb Erekat, the former Palestinian chief peace negotiator said, "Jerusalem is not off the negotiations table, rather the U.S. is outside the international consensus" (Khoury 2018). It is worth mentioning that Trump's Jerusalem announcement was uniformly opposed by 128 UN states and received a worldwide condemnation, especially from the Palestinian leadership.

This controversial move was characterized by Dr. Hanan Ashrawi, PLO Official, as illegal, absolutely irresponsible, and in direct conflict with the foundational tenets and imperatives of the peace process. In an interview with BBC World News, 2017, she described this announcement as "extremely dangerous and extremely discouraging" as is detrimental to the overall stability and security of the region and would further scuffle any chances for real peace. Recognizing the colossal failure of the so called "peace process," which she attributes to the U.S. bias towards Israel and Israel's impunity and complete disregard of international humanitarian laws among other things, she makes the point that by doing so, Trump's announcement cannot be looked at as a positive step towards peace and calm, but rather an incitement of violence and conflict in the region considering Jerusalem's stature and its religious, spiritual and cultural sentiments for the Palestinians. She voiced the overwhelming Palestinian position on the issue by affirming that Jerusalem is a "Palestinian city" that "belongs to the Palestinians" and that "history cannot be turned into forgery" (Ashrawi 2017).

As a way to protest this consequential decision, the PLO leadership responded by boycotting Mike Pence's visit to Israel, describing his speech to the Israeli Knesset as an attempt to impose his "messianic zeal" and "biblical homily," in the words of Dr. Hanan Ashrawi. She astutely contends that "It's Pence who is trying to reshape the world the way it was 3000 years ago … He thinks he can superimpose his messianic zeal … a biblical homily that he delivered at the Knessent, to say that the Bible applies to 21st century in a very literalist fashion. This is not the way to make peace … if you want to make peace, then you adhere to international law" (Ashrawi 2018). Alluding to his so called "deal of the century," Trump threatened to pull aid to Palestinians if they do not pursue peace negotiation with Israel and Nikki Haley, the U.S. Ambassador to the United Nations, blasted the Palestinian President Mahmoud Abbas for "lacking courage and the will to seek peace" (Khoury 2018). The Palestinian leadership, in the words of Saeb Erekat, responded that "if Jerusalem is off the negotiation table then the U.S. no longer has a seat at that table." The Head of the PLO General Delegation to the U.S., Husam Zomlot, in the same vein, was also very critical of this announcement stating that Trump did not take the issue of Jerusalem off the table as he claims; he "took the table altogether," in reference to the negotiation table (attributing the failure of the peace negotiation to these sorts of ill-advised and one-sided political maneuvers) (Wright 2018).

Third, it solidifies the already known fact about the U.S. impartiality towards the conflict and its unequivocal support of Israel. This U.S. position also shows extreme bias in favor of Israel and an accommodationist stance for the illegal occupation and annexation of Jerusalem. Dr. Ashrawi, critical of this announcement, pointed out that "recognizing reality" instead of basing this recognition on international law does not necessarily means the right thing to do, and in this context it is dangerously detrimental (Ashrawi 2017). She, akin to many others who made strides towards promoting and legitimizing the language of peace and reconciliation, sees no way out of this impasse other than taking Israel to the International Criminal Court to hold it accountable for its violations and continued disregard of international law. The U.S. embassy plans to be moved to Jerusalem on the same day marking the establishment of Israel seems insensitive to the Palestinians considering that this day happens to be also the day that the Palestinians regards as the "Nakba" or catastrophe. This decision left no doubts in the minds of those who still question the neutrality of the U.S. as a mediator in the peace process between Israel and Palestine. As a response to the planned U.S. embassy move to Jerusalem, the Palestinians in the Gaza Strip organized the Great March for Return. Akin to the Boycott, Divest, and Sanction (BDS) movement, the goal of these efforts is to put pressure on Israel to abide by international law, lift the Gaza blockade (now in its twelfth year), resolve the issue of Palestinian refugees based on the 194 UN Resolution, and put an end to the occupation of Palestine. These are examples of nonviolent resistance that the Palestinians launch to voice their disappointment and frustration about how the diplomatic effort has utterly failed them and exacerbated their situation. Now, they are further from achieving peace than ever. A fundamental change to their current situation is badly warranted, including lifting the siege on Gaza, halting settlement expansion, and addressing the issue of Palestinian sovereignty and the right of return, to name a few. Otherwise, we will continue to see these types of non-violent measures being deployed by the Palestinians in the future. Significant in this context is the unescapable impact of Trump's deal of the century on the political map of the Middle East and the future of the Palestinian-Israeli diplomatic relationship, which remains to be seen. This is where things stand today on the diplomatic front between the two parties.

4 What Went Wrong?

The failure of previous diplomatic attempts to land sustainable and robust peace agreements is primarily a byproduct of a number of deficiencies that these negotiations suffered from. It is crucial that key lessons and parallels can be extrapolated from these diplomatic attempts that took place in the past to guide future negotiations to ensure their success. Although suffered from many pitfalls, major issues can be extracted so lessons can be learned. These major pitfalls are related to the bilateral agreements used and the Palestinian representation, lack of confidence building measures (CBMs), the impact of power-structure imbalance, settlement expansion, international law and the perceptions of "pre-conditions," peace versus process, the perceived failure and lack of support on the local level, applying a technical solution to an adaptive challenge, transformative versus transactional negotiations, water was excluded or negotiated separately, and the impartiality of the U.S. as a mediator. Although discussed in this chapter as we followed the chain of events that conspired, I will talk briefly about each of these important issues to give them more emphasis.

First: Bilateral Agreements and the Palestinian Representation

It is pivotal that all affected parties be identified and involved in direct negotiations. Otherwise, by excluding a key stakeholder, such as the Palestinians, a plethora of issues will challenge the negotiation process and any resulting agreements due to the *ex parte* syndrome. It is no secret that Israel has always opposed and abhorred the idea of contacting or negotiating with the PLO, a position that was bequeathed to the U.S. administration to adamantly implement until the early 1990s when Israel itself was forced to abandon this position. From the early stages of the negotiations, the issue of the Palestinian representation constituted a major challenge, as the PLO was not recognized by either Israel or the United States.

This constant Israel's obduracy and full-throated opposition to any negotiation with the PLO as a formal partner was indicative of the larger geopolitical occultation that Israel was forcing the PLO into for a very long time even prior to the so called "peace process." Israel's intransigent position towards the Palestinian independence, and by extension their aspiration for statehood explains why Israel spurned the idea of direct talks with the PLO. This caused a diplomatic hurdle that inhibited the Palestinian ability to be a part of any negotiation involving Israel, including those that Israel conducted with other Arab states, such as Jordan and earlier with Egypt. It was precisely this Israeli denial and refusal to acknowledge the Palestinian rights for independence and sovereignty that flawed the Oslo process, which was by design meant to postpone and deflect these rights rather than to address them in a meaningful way (Said 2003). The willfully deceptive rubric of the "peace process" served as an assiduous diplomatic protection to cover, reinforce and legitimize the Israeli occupation and dominance over Palestinians—a downtrodden and dispossessed people. Due to this Israeli position, along with the lack of coordination and alignment between the different negotiation tracks, the Palestinians were represented as part of the Jordanian delegation. King Hussein's decision to disengage from the Palestinian West Bank, as well as the joint Jordanian-Palestinian delegation, raised a problematic question regarding the Palestinian representation. In all of the preceding bilateral diplomatic fronts led by Jordan and Israel, which successfully produced the Israeli-Jordanian Peace Treaty of 1994, the Palestinians were almost entirely absent. This personifies a case of "negotiation without representation," which is indicative of the Israeli position towards the Palestinians and can be a predictive measure for what to expect as the outcome ceiling for the Palestinian independence.

Owing to the fact that Israel and Jordan precipitated in such serious talks for a final settlement, while the Palestinians and Syrians were left out of these diplomatic arrangements, these ghastly negotiations were unable to produce a fair outcome for any of the parties involved any way you look at it. Given this blatant pitfall, it is not surprising to see the overwhelming skepticism of the majority of the Palestinian community who utterly spurned the outcome of these negotiations from which they were excluded. If Israel is not willing to give the Palestinians the autonomy to negotiate as an independent party, how could they possibly allow them to have an independent state of their own as a result of these negotiations? They could not. And how could the Palestinians possibly think of any other outcome of such negotiations? Given these circumstances, the stillborn outcome was predictable and almost guaranteed to lead to what eventually became a harsh, yet unescapable, reality for the Palestinians to swallow from the very onset of these negotiations.

Israel always advocated for bilateral, rather than multilateral, diplomatic engagements to avoid involving the PLO, which became one of the foremost issues to challenge the negotiation process. Bilateral agreements are historically more common than multilateral agreements due in part to the fact that they are relatively easier to negotiate and faster to ratify and govern (Wolf 1998). Although Israel's tendency to advocate for bilateral agreements may hinder the ability to attain integrated river basin-wide management, this bilateral negotiation tactic employed by Israel gave it greater advantage to be able to isolate parties and further control the different negotiation tracks, while keeping the Palestinians out of them (Phillips 2008). By doing so, Israel was at the center of the negotiation and was successful in evading coalition formation that could pose a real threat to its interests and negotiation agenda.

Second: Lack of Confidence Building Measures (CBMs)

The prospects for generating and evaluating acceptable options would be significantly enhanced with the adoption of sensible confidence building measures during the negotiation, or better yet before the process begins during the pre-negotiation stage. Parties were sporadically engaged in negotiations that can only be described as "diplomacy on the run." They attempted resolving specific, and sometimes complex issues, but without a comprehensive agenda or long-term negotiation strategy to foster trust and ongoing collaboration. Because the parties were eager to resolve the conflict for different reasons, the outcome was rushed and lacked focus. Despite the fact that the parties were in a state of war when they decided to enter into these negotiations, they made little effort to build trust and confidence to eliminate sources of suspicion and anxiety. Instead, they pressed on with their negotiations and the outcome was more conflict and tension during and after these negotiations. Considering the history of this conflict and the emotional dimension involved, it is pivotal that the parties demonstrate a great deal of goodwill and trust prior to engaging in any negotiations. The fact that they primarily focused on ethnocentric and entitlement issues resulted in a situation of talk-fight, where both parties found themselves on a collision course with one another. This tension and mistrust grew as the negotiation progressed and the implementation ipso facto suffered because of that. This particular aspect was ultimately a key contributing factor that led to the demise of the "peace process."

Third: Impact of Power-Structure Imbalance

Hydro-hegemony is the advantage that a state can leverage as a result of a number of factors including its military, political, economic, or hydro-strategic position (as an upstream riparian, which usually lends it a significant advantage over other downstream riparians). It is also related to the state's water exploitation capability including infrastructure development and technological advancement. The impact of power-structure imbalance (or *hydro-hegemony*) has been evident in all of these diplomatic undertakings, which tipped the scale towards the side of Israel (being the most powerful state). This created an uneven playing field where Israel was able to shape and choreograph not only the outcomes of the negotiation but also the process by which these outcomes were reached. As a result of its relatively strong *hydro-hegemonic* position, Israel's sulfurous attitude towards the Palestinians and by extension the PLO has always reflected its reluctance to negotiate. This is because cooperation offers little or no benefit to the most powerful state, due in part to the potential of relinquishing its most favorable position, coupled with the fact that Israel's BATNA (best alternative to a negotiated agreement) is synonymous to the status quo that Israel created and is keen to maintain. In this regard, Israel has little to gain and potentially a lot to lose as a result of a fair deal that recognizes the right of the Palestinians to their land and water resources both of which are currently under Israeli control. As such, when it finally acquiesced to negotiate with the PLO, Israel (with the backing of the U.S.) was dictating the rules of the game and confined its cooperation to specific issues in order to maintain its dominance. Israel was able to gain the upper hand regarding which issues to negotiate, to exclude, or defer to the final status negotiation. Critical to this context is the emphasis on maintaining Israel's national security and eliminating hostility in the region, which was the main focus for Israel. Moreover, to again and maintain this advantage as a dominate player in this lopsided diplomatic endeavor, Israel was not only able to extract considerably significant amounts of water compared to the Palestinians, but it was also able to manipulate its water consumption data to conceal the reality on the ground. Israel also manipulated the narrative of these negotiations and their outcomes through its public discourse to build consensus over a singular form of "truth" about its water consumption, actual water needs and water availability. The outcome was also reflective of these power dynamics that led to coercive relationship, flimsy agreements, and the political and diplomatic straitjacket within which the Palestinians have found themselves ever since they accepted these agreements. More attention should be given to neutralizing the impact of power-structure imbalance in order to level the playing field and offer an equitable platform for all negotiating parties to lead constructive negotiation and help in moving the conflict from intractability to cooperative discourse.

Fourth: Settlement Expansion, International Law and the Perceptions of "Preconditions"

Evidence suggests that these talks were not concerned with peripheral administrative or technical issues. Their main focus was to establish an understanding to bring an end to the conflict by addressing and eliminating its main root causes rather than the symptoms and resolving key hot issues, such as land, borders, sovereignty, the right of return, and Jerusalem. From a Palestinian standpoint, the issue of Israel's settlements was not addressed directly in these talks because it was presumed and understood by everyone on both sides of the isle that those settlements are illegal under international law and need to be dismantled. As such, they were not part of the issues formally considered in the negotiation. Some, however, argue that this was a deliberate strategic move on part of Israel to avoid discussing this hot issue which gave Israel the cover it needed to expand its settlements without any opposition, ramifications, or accountability. In other words, the whole peace process was hinged on a *nudum pactum* of a sovereign Palestinian state inside the 1967 boundaries and the assumption that any further illegal Israeli settlement activities would be scuttled and completely dismantled. This, however, was never accomplished, and is currently described as an illusion at best, namely that these negotiations were conducted with a high level of secrecy and ended up focusing on procedural and administrative issues to run a makeshift Palestinian government without real sovereignty, independence or authority. The PLO was in a way pressured to accept this arrangement in lieu of a comprehensive agreement that would provide guarantees to the Palestinians of an independent state of their own. The PLO made significant concessions in recognizing Israel as a state but failed to secure the same recognition from Israel for a future Palestinian state. The PLO leadership was in a way preoccupied by its own agenda of gaining recognition from Israel and the international community as the legitimate representation of the Palestinians, which made them lose sight of the important issue of gaining recognition of the Palestinian right to statehood. This was a major pitfall of these negotiations. In doing so, the PLO was not accountable to, and further misrepresented and abandoned, the Palestinian people who viewed its leaders as legitimate representation of their national struggle and right to co-exist in peace and harmony with their neighbors. Moreover, the premature Palestinian negotiators' reliance on international laws without first securing acknowledgment or acceptance from Israel created an unbridgeable gap in relation to creating mutual ground rules of engagement to govern their interaction during and after negotiation. These main issues mentioned earlier in this chapter were not addressed or resolved in any meaningful fashion and continued to this day to be major obstacles to achieving real peace.

While the two sides were negotiating over land and sovereignty, one side was expanding its control of the very land they were negotiating over. Needless to say, this galling and counterproductive practice, coupled with Israel's obduracy in allowing this issue to even be a part of any discussions with either the American sponsor or their Palestinian counterparts, was evidently detrimental to the negotiation process and its outcome to say the least. Equally galling, was the U.S. position which was clear—there was no "magic wand" that the U.S. can utilize to force Israel to freeze its settlement activities while the two sides are negotiating, as explained by the U.S. Secretary of State James Baker. Although the U.S. enunciated its disapproval of Israel's settlement expansion on many occasions throughout the different U.S. administrations, its effort came to a naught as it utterly failed to exercise any pressure on Israel to actually dissuade settlement construction and land appropriation. Instead, during these negotiations, Israel continued to booster settlement building driven by its ever insatiable territorial appetite. By so doing, the U.S., not holding itself accountable to its promises to the Palestinians or its role as the sponsor of the peace process, indirectly heralded a subtle "green light" for Israel to continue its construction of settlements on Palestinian land notwithstanding its illegality. The issue was made even worse as these Palestinian demands to halt further settlement expansions in the West Bank and East Jerusalem were cast a side by both Israel and the U.S. as unwarranted "pre-conditions" to entering negotiations.

Moreover, this Israeli practice foundered the chances of achieving a "two-state solution," which has evidently become a "two state delusion," considering the current situation on the ground that was made possible by the unyielding Israeli insistence on continuing its settlement expansions and the U.S. complacency or complicity in this regard. In a nutshell, the so called "peace process" has been effectively utilized as a veneer cover to reinforce Israel's grip on occupied land, expand its settlement construction while attempting to pacify the Palestinians and their leadership with a façade self-governing arrangement that effectively serves as Israel's police in the West Bank and the complete isolation of the Gaza Strip. This issue of the Israeli settlements remains to be seen as one of the major irreconcilable differences between the two sides, although the international law is abundantly clear about its illegality. In future negotiations, these important aspects must not be perceived as "pre-conditions," but rather legal obligations that Israel must abide by prior to entering any negotiations.

Fifth: Peace Versus Process

Reaching a resolution is important, but reaching a resolution that's acceptable by the parties involved in the dispute is even more important. The process by which this outcome is reached is as crucial as the acceptability of its outcome. However, in the negotiations between Israel and Palestine, the procedural aspects dominated the substance and the effort was primarily focused on the process not the outcome. The so called "peace process" was thought of as a means to an end—a process that would ultimately lead to peace ending years of hostility and bloodshed; a two-state solution where the Palestinians would have independence and self-determination. The overarching objective of this peace process that was ostensibly directed and sponsored by the United States was to reach a just, lasting, and comprehensive peace. However, in reality it was just simply a process without real outcome. The process of peace talks itself has become an objective, in and of itself, which gave it a life of its own without real substance or actionable outcomes. The two sides became complacent with the idea that establishing peace talks by meeting and negotiating is synonymous to achieving peace, regardless of what actually happens or does not happen in reality. The peace process did not even intend to put an end to the ongoing Palestinians misery and harsh reality but rather to make it look slightly less unbearable (Said 2003). The process instead failed to achieve this nadir goal by exacerbating the immiseration of the Palestinian communities and further contributed to its denial. According to Said, 2003, it was pacification not peace that was sought at Oslo.

What the so called peace process did was devastating for the overall stability of the region and the Palestinian aspiration of having their own free and independent state. While these negotiations were taking place, Israel was actively changing the reality on the ground by wreaking havoc on the occupied Palestinian communities in the West Bank and Gaza. Israel took advantage of these years of negotiations to deny the Palestinians their rights, impose more restrictions, confiscate more land for its illegal settlements and advance its position and overall military control over these territories. This prolonged peace process allowed Israel to apply significant pressure on the Palestinian Authority to crush any opposition or revolution against Israel or the process of negotiation itself. By continuing to engage in these talks with Israel, the Palestinian Authority found itself doing the dirty job for Israel by acting as Israel's police in the occupied territories to maintain calm and suppress any opposition. Under the guise of peace talks, Israel was able to enable and legitimize its illegal actions in the occupied territories and gain political impunity through negotiation to reinforce the *status quo* of an occupying power.

Sixth: Perceived Failure and Lack of Support on the Local Level

Public perception of representation and fairness is a critical element in the level of acceptance and implementation success of the outcome of any negotiated settlement. Acceptance and support of the negotiation outcome is primarily contingent upon the general perception of the negotiation process. Of gargantuan importance is the public perception on both sides of the border, which has not been adequately addressed and as such emerged as unfavorable and unsupportive. The psychological impact of changing the course of the dispute from incitements and hatred to a sudden cooperative arrangement (without a parallel public awareness campaign) was detrimental in this context. Perception of the general public is primarily shaped by the history of the conflict, their knowledge of the agreement, the media coverage, and the inherited perception of the "other." Considering the intractability of the conflict and its history and the prescription of each other, neither nation was psychologically ready to undertake direct negotiation with the other party. This was evident in the overwhelming public reaction on both sides that manifested itself in certain violent acts in response to this alternative reality (of cooperation) that was prematurely imposed on the two nations. For the Israelis, this was an abandonment of their leaders to their intransigent position towards Palestinian independence. For the Palestinians, this was perceived as a western panacea and occultation of the refugees' right of return stipulated by the United Nations General Assembly Resolution 194, and by extension their right to their land and independence. The separate (but related) assassination incidents of Israeli Prime Minister Yitzhak Rabin and Egyptian President Anwar Sadat were a perfect illustration of such violent and extreme reactions attempting to scuffle this new and difficult reality for many (Linskey and Heifitz 2002). Israeli Prime Minister Yitzhak Rabin was assassinated by one of his own hardliners, the very same people who bought into his own radical views that he presumably abandoned by engaging with the Palestinians in peace negotiations they were not ready for. This incident exposes the level of hatred that was fueled by decades of incitement from their very own leaders who advocated for many years for violence against the Palestinians during the Intifada and beyond. Such extreme reactions personify a widespread rejection of these agreements on the local level and a perception of a colossal failure associated with these diplomatic efforts.

Aside from perception, reality also speaks volumes in this regard. Almost twenty five years of negotiations between Israel and the PLO resulted in more bloodshed and conflict than ever. The Palestinian Authority was intended to be an embryo for the establishment of a larger Palestinian state, which never took place. Contrary to popular belief, the Palestinians now are actually much farther away from achieving their statehood and independence as a result of this failed peace process. The Palestinian Authority personifies a symbol of failed politics and a result of a process that is viewed as nothing short of a disaster for the Palestinians. Feeling the direct impact of the "peace process," security cooperation with Israel is one of the most hated aspect of the whole relation, whereby the Palestinian Authority has been effectively acting as Israel's police in the territories without real authority or autonomy. It is only a façade behind which lies no prospect or possibility of a Palestinian state. To this day, the term "security cooperation" which was originally intended to maintain the implementation of joint cooperative agreements between the PLO and Israel has become probably the most loathed act among the Palestinian community and is perceived as a dirty business almost as unpatriotic and unforgivable as treason. As a result, both the Israeli and PLO leadership lost political support, credibility and legitimacy in the eyes of their respective constituents who viewed the whole process as a political farce and a western panacea to legitimize Israel's occupation and produce a situation that can only be described as a political and diplomatic debacle.

Seventh: Applying a Technical Solution to an Adaptive Challenge

Needless to say, the Palestinian-Israeli conflict, albeit seemingly soluble, is extremely nuanced and epitomizes an adaptive challenge. It is exceedingly complex, multidimensional, highly contentious, and has many moving parts and parties. Adaptive challenges are systematic, infused, changeable, hard to diagnose and easy to deny and ignore. If overlooked or ignored, however, adaptive challenges can result in dire and consequential ramifications. A technical solution cannot provide adequate response to adaptive problems, as they are concerned with belief systems, behavioral and cultural changes, and relationship building. These challenges require examination of the current situation, collaboratively identifying desired future conditions and institutionalizing and operationalizing a plan to help the parties realize this desired future and close the gap.

In all of these previous diplomatic efforts, the negotiations were mainly concerned with technical issues and were for the most part conducted on a technical level. For example, the pre-negotiation process over water allocation between Israel and Jordan, which started years before it was actually announced to the public, was purely technical in nature. It began in the form of "Picnic Table Talks" to determine and manage the day-to-day hydrologic operations of the Yarmouk River. The parties involved in the talks were mainly technocrats, scientists and engineers to form a joint fact finding team (JFF). The goal of this technical team was to share hydrologic data and discuss technical solutions regarding water allocation along the Jordan River valley. The final agreements were not binding and reflected this technical nature of the discussion and was only concerned with technical solutions to the water shortage problem. It eventually produced *ad hoc* technical agreements to manage the timing and volumes of water distribution, without addressing the core issues of the conflict which are much more nuanced and wicked. Although these technically-driven Picnic Table Talks provided a factual basis for subsequent political negotiations and eventually the 1994 Peace Treaty, it did little to change people's perceptions or building trust. This is evident by the fact that the nature of the current relation between the two countries can hardly be characterized as one of mutual trust. This problem became more pronounced in the 1999 drought that led to finger pointing and a great political drama as the parities failed miserly in managing the implementation of the treaty in a meaningful and sustainable way. What was needed in these talks was a multi-scalar, transformative negotiation platform that offers a great deal of resilience to combat and respond to these adaptive problems.

Eight: Transformative Versus Transactional Negotiations

A major flaw encountered in these negotiations that seems to have led to only fragile and unsustainable agreements at best is related to the position-based approach that attempted to address only the question of entitlement to land and water resources instead of addressing the actual needs of both sides. This practice is reflective of the classic process of distributive (transactional) bargaining, where one party's gain, while viewed as winning, is considered a loss for the other party. This problem was compounded by the fact that the Zone of Possible Agreements (ZOPA) was not even attainable, which led to the tendency to *divide* the shared resources rather than to *share* them (Scheumann 1998).

Transactional negotiations are generally beneficial to help the parties focus on trading across value points as long as all parties are on the same page regarding the negotiation agenda and priorities. However, when the parties are involved in an emotional and intractable conflict, such as this one, where each party views the other party's win as its own loss, and the outcome of the process is unclear, this transitional negotiation style breaks down and becomes dysfunctional. By contrast, transformational negotiations provide a device for the parties to enable them to collaboratively identify and prioritize issues of perceived high value to each of them, establish ground rules regarding engagement mechanisms, and jointly create different value streams to consider, assess, and trade across. Considering the state, magnitude, and nuanced nature of this intractable dispute that the Palestinian-Israeli conflict personifies, a transformative negotiation style is more conducive and almost warranted to achieve the aforementioned benefits from the ability to collaboratively augment and clime the value chain. A transformative framework of negotiation will be discussed in detail in the next chapter.

Almost all of these previous diplomatic efforts represented transactional negotiations, due in part to Israel's obduracy in accepting the PLO (and by extension the Palestinians) as an equal partner in peace. This, coupled with the U.S. bias and lack of focus on collaboratively finding a common purpose and a transformative agenda, put the prospects of finding workable and pragmatic solutions in jeopardy and created a strategic blunder instead. For instance, the declaration of independence that was ceremonially signed on the White House lawn under the aegis of the American mediation was designed in a specific way to legitimize Israel and make it near impossible for the international community to criticize its actions or exercise any type of pressure to end its illegal practices in the Palestinian Territories. The Madrid Talks were strictly based on the guiding principles of the international law, including the UN resolutions 242 and 338, which called for their implementation "under appropriate auspices." Both resolutions clearly outlined a set of criteria and guiding principles to govern the relationship between the Israelis and Palestinians. The "land for peace" formula was the heart and soul of these negotiations, which everyone understood to be the overarching goal. As such, this "land for peace" formula represented a classical transactional approach to negotiation. While it created some focus for the teams to hone in on during their negotiations, it also created a serious diversion from the main issues that are key to the final resolution, whatever that might be. In addition, under this negotiation style, the parties seemed to have skipped a major step in their negotiations and missed out on the opportunity to identify and prioritize issues of high importance to each of them, and therefore, failed to collaboratively search for ways to augment and capture greater value streams. The transactional nature of the talks was reflected in the type and quality of the resolutions that were proposed and put on the table for considerations and ultimately led the process to a deadened due in part to the irreconcilable differences that were not addressed directly. This is because the transactional nature of the engagement skewed the agenda towards tactical items that were only concerned with transactional issues, rather than a transformative approach in order to bridge their differences and elevate the current stagnant and intractable situation to a more amicable one. These differences did not subside as the parties, finding themselves in a *par delictum* situation, carried on with their negotiations paying no or little attention to eliminating, or at least neutralizing, these differences. Instead, they persisted, festered, and were even exacerbated by the transactional nature of the interaction that obfuscated the crux of the issue at hand, and induced more clashes and eventually brought the whole process to a stalemate.

Another fundamental flaw that made these transactional negotiations even more challenging is related to the lack of accurate data reading actual water needs that are ubiquitously agreed upon. Utilizing authoritative and representative data in transactional negotiations is crucial. The data regarding water consumption, water availability and current and future demand was either exaggerated or underestimated reflective of the party's motives to gain more access and undermine that of the other party. This had a direct impact on creating a gap between reaching agreements and implementing them. Almost all of the previous negotiations suffered from the lack of implementation. As a result, this process that was once thought to be revolutionary is now yielding another surprise. These agreements generated clashes during implementation and ultimately reverted the process back to a "talk-fight" style that brought the negotiation between the Palestinians and Israelis to a screeching halt.

Ninth: Water was Excluded or Negotiated Separately

In many cases, water was not addressed, talked about or considered, and in other rare cases was negotiated in isolation of other tangible topics. For the most part, water was one of the major issues that was deferred to the final round of negotiations between the two sides, and still constitutes a major obstacle to achieving peace to this present day. Although water was negotiated both in the bilateral track and one of the multilateral groups, the issue of equitable allocation of water was never adequately addressed in a meaningful way (Elmusa 1995). Serious resolutions over water resources allocation were not part of these attempts due to the fact that water was one of the main three issues that were deferred to the final rounds of negotiations. The political and resource issues, reflecting the realms of "high" and "low" politics respectively, have been always dealt with in separation from one another. The evident mismanagement of the negotiation process lifecycle that led to this pronounced separation and lack of an integrative approach to jointly resolving these intertwined issues is regarded by many to be a fundamental reason for the failure of these previous diplomatic initiatives (Daoudy 2009; Phillips 2008). For instance, there has been many documented diplomatic attempts that tended to strictly address water resources allocation only, such as the Johnston Negotiations of the 1950s, the "water-for-peace" negotiations in the late 1960s, negotiations over the Yarmouk River in the 1970s and 1980s, and the Global Water Summit Initiative of 1991 (Wolf and Newton 2008). These attempts, which had the potential of generating cross-track synergy to leverage the advantage of bundling multiple issues together, failed in settling the conflict as they were conducted almost entirely in separation from the larger political discourse. Given its cross-boundary nature that does not recognize artificial political divides, the water issue cannot be resolved alone without comprehensively looking at the overall political and socio-economic discourses in their entirety. This was one of the major befalls in these previous negotiation attempts that both sides of the isle need to be cognizant of in the next rounds of negotiations, to ensure alignment, maximize synergy, and leverage cross-issue advantages that would potentially and effectively allow them to trade across different value streams.

Tenth: Impartiality of the Mediator

The U.S has been involved as a mediator and a broker of peace between Israel and Palestine. This, however, was proven problematic throughout the negotiations that ever took place between the parties, considering the strong American-Israeli relationship that made it nearly impossible for the U.S. to act in an impartial fashion. Since the time of President Truman to this present day, the U.S. policy towards Palestine which has been greatly influenced and sometimes shaped by the Israeli lobby, the American Israel Public Affairs Committee (AIPAC), has rarely changed. It is one of the most ironic and troubling aspects of the entire so called "peace process" that the U.S. has been serving as a mediator and a peace broker given its unwavering support of its favorite ally Israel. The U.S. in many occasions acted to serve Israel's interests, namely through its diplomatic influence and intervention at the United Nations on behalf of Israel—in a way acting as Israel's lawyer to block any attempt to condemn or hold Israel accountable for any of its actions, particularly the expansion of its illegal settlements on Palestinian land and its documented human rights violations against the Palestinians. When it comes to its foreign policies, the U.S. unequivocally demonstrated its partiality towards and complete support of Israel. This has consistently been demonstrated with the U.S. relentless effort to block any

UN Security Council resolution that would condemn Israel's actions against the Palestinians even when it is supported by the majority of the global community. This U.S. policy has been so pervasive and predictable that one can easily anticipate this standard U.S. response which regardless of the issue at hand is ready and guaranteed to thwart any international attempts to blame or condemn Israel for any of its actions. This has undoubtedly compromised the U.S. global status and contributed to the widespread perception that the U.S. has been acting as Israel's lawyer in all of its previous diplomatic engagements (Khalidi 2013). Not surprisingly, the peace process evidently failed to create a cadence of accountability by holding the U.S. to its promises and assurances made to the Palestinians in the Letter of Invitation. This contributed to the overall impartiality of the process and its outcome that has produced a progressively and disastrously deteriorating reality for the Palestinians who found themselves in a diplomatic and legal blunder.

The most recent development and escalating protests on the Gaza border fence that left many dozens killed and thousands injured following the U.S. decision to move its embassy from Tel Aviv to Jerusalem signaled a clear message to the international community and to the Palestinians that the U.S. is on the side of Israel and can no longer be trusted to broker future peace. The images of Ivanka Trump's opening ceremony with shining smiles next to bloody images from the Gaza border of kids being shot by Israeli snipers just fifty miles away from Jerusalem signaled a stark contrast of two very different worlds and brought back memories of colonial British sickening elitism showing no regards to human lives and suffering. This was further exacerbated by unexperienced Kushner's remarks that blamed the victim by criticizing these demonstrations as "part of the problem and not part of the solution" instead of recognizing them for what they are—peaceful demonstrations to demand equal rights and freedom. This was deeply troubling. Equally troubling was the timing of the U.S. celebration of the move of the embassy one day prior to the *Nakba* which commemorates the forced exile and displacement of the Palestinian communities as a result of the founding of Israel in 1948, which added insult to injury. Although widely criticized for its lethal response, Israel with the help of the U.S. justified its actions as simply "defending itself." Many, including the U.S. very close allies, considered this to be the last blow to the peace process as this move not only compromised the prospect of any current or future peace talks, but it also compromised the U.S. position as a potential mediator in any future diplomatic attempts that involve Israel. Given its tight relationship with Israel and track record in the previous diplomatic efforts, the legitimacy of the U.S. as a neutral mediator has been compromised, and as such, it can no longer play the role of a good faith mediator in any future negotiations. Unless a serious and demonstrable change in its foreign policy pertaining to Israel is achieved, any future U.S. intervention will be viewed as biased.

These lessons are critical components not only for understanding the current situation and where things are now, but also and more importantly guiding future relationship and any peace making undertaking between the two sides. These shortcomings from which previous negotiations suffered can be used as a platform to educate both sides on important lessons to be learned in order to articulate a mutually agreed upon vision for peace and reconciliation for the sake of resurrecting a lasting peace settlement that can adequately satisfy the interests and expectations of both sides. The final chapter will provide a path forward that, while grounded in the lessons learned from these previous mistakes, is capable of charting the course to a more practical and pragmatic roadmap for a peaceful settlement.

Chapter 5
A Pragmatic Conflict Resolution Framework

Justice is the essence of peace.

—Ali Abunimah

In my previous book I provided an agenda and a framework for a transformative approach to conflict resolution. The raison detre for this chapter is to provide an example for the adaptive application and utilization of this transformative framework in the context of the Palestinian-Israeli conflict to transform water negotiation from a non-cooperative to cooperative state, where both sides can engender equitable hydro-diplomacy and satisfactory outcomes. In this chapter, I will illustrate, as dispassionately as I can, how to use this practical framework to move forward with solid recommendations for a real resolution that would cater to the very unique, yet attainable needs of both nations.

By operationalizing this transformative framework that was developed in my previous book, this chapter will provide guidance for the future negotiation between Israel and Palestine. It will utilize these best practices and proposed conflict resolution *discipline* to advance agreements towards a more robust and sustainable settlement that would guarantee its acceptability and successful implementation. This theoretical framework will constitute the foundational knowledge and building blocks for a more pragmatic approach and a roadmap for how negotiation should be conducted, who should be involved, issues to be discussed and criteria for acceptable and equitable outcomes (with water as a bonding factor and a bridge to peace).

This chapter aims to bring the topics discussed in the book into a neat and tidy punctuation with concluding remarks that provide strategic, procedural and execution-related aspects and next steps. It will briefly reiterate the shortcomings of the previous negotiations to learn from these mistakes and provide a go-forward strategy for negotiation including Israel's obligations to abide by international law (which was misunderstood by previous negotiators and was mistakenly perceived as unnecessary preconditions to negotiation). These are not preconditions but rather obligations and prerequisites for Israel to halt all illegal settlement activities and expansions on Palestinian lands, water allocation equity attainability, right of return and compensation for the Palestinian refugees, and ending the illegal siege of Gaza. These are essential steps that need to be accomplished prior to any negotiations to ensure their success in yielding tangible and acceptable outcomes. This was not accomplished in the past and constituted a major point of contention and therefore undermined the success of these negotiation attempts. These and other factors will be outlined and presented for future negotiation to be fruitful and productive resulting in a lasting peace for all.

A. Abukhater, *Palestine - Peace by Piece*,
https://doi.org/10.1007/978-3-030-03985-1_5

1 Current Situation: Everything Has Changed but Nothing Is Different

There are a multitude of thorny issues surrounding the question of Palestine and the Palestinians. The bristlingly complex nature of this conflict dwarfs any solution regardless of its magnitude. A more comprehensive one that addresses the needs and aspirations of both peoples is disparately needed to put an end to this conflict once and for all, and bring sustainable peace and stability to the region. In the wake of the Oslo agreement, peace was thought to be possible between Palestine and Israel and the state of violence was seen as a thing of the past. Things were looking up and people on both sides expected real changes towards a more peaceful settlement to the conflict. Twenty five years later, change did in fact happen but for the worse. And for the Palestinians little has changed, with the exception of more land appropriation and less freedom. Gaza, characterized by many as an open-air prison, has been under tight and deepening siege since Hamas took control in 2007. In the West Bank, a record number of Jewish settlement units have been built on Palestinian land (Blumenthal 2015).

For the average Palestinian, and especially those who truly believed in peace as a realistic possibility to put an end to this conflict, things became crystal clear with every passing day—more settlements, more associations and more siege. In 2008/2009, Israel launched an offensive on the civilian population of the besieged Gaza Strip. The result was the killing of over 1,400 Palestinian civilians and the devastation of Gaza's major infrastructure. The scenario repeated itself in November of 2012, when Israel decided to unleash its war machine on the besieged population of the coastal enclave. And with the recent Israel aggression in 2014 aimed to destroy what's remaining of Gaza's infrastructure and any elements that could potentially constitute a viable Palestinian state, the prevalence of a benign sense of peace has been completely and decisively dismantled. For many people, the dream of a peaceful settlement is now shattered. Amending this broken promise of peaceful existence might be too much to hope for under the current circumstances as long as the siege and closure of Gaza's border is maintained and enforced.

The bitter division between the Palestinians in the West Bank and Gaza is a self-inflected wound that exacerbated an already intractable situation and made things much more intolerable for the people on the ground. In fact, many people in Gaza have not been paid for many months and years despite their disciplined insistence to go to work every day. This division, which is reflective of a dangerously broken bond and out of kilter dynamics in the Palestinian house was not a coincidence. It was a byproduct of internal and external intervening forces that benefited from the division of the Palestinian leadership effectively creating two separate governments to rule the two noncontagious territories—one in Gaza and the other in the West Bank. Gaza in particular suffered the brunt of this divide. And above all, there has been a mutual detriment to both Palestinian territories' ability to create, in any meaningful fashion, a united political front towards self-determination. It goes without saying that it is in the best interest of the Palestinians that they make a conscious decision to eliminate and uproot this internal division whenever it rears its ugly head. Moreover, unless Israel lifts the siege and allows the unconditional rebuilding of Gaza, halts and completely dismantles its illegal settlements in the West Bank, and allows the Palestinians to have their own state, peace and security is not going to be in reach anytime soon.

2 Where to Go from Here? Water as a Bridge to Peace

In the Middle East, water and war are often discussed in the same context. The tight connection between water and other geopolitical issues drove the desire to acquire undivided and exclusive mastery of land, and by extension water being the region's most precious natural resource. In this region, with its weighty political struggle, water is perceived not only as a precious natural resource, but also as the sustainer of life and peace. Given its pivotal role in shaping the geopolitical landscape, water can be a point of contention, severe political conflict and vicious disagreement, especially when basic human needs are not met. In this semi-arid region, water is an extremely confined and precious natural resource that is believed to have exacerbated an already smoldering and protracted conflict.

The significance of water fueled many conflicts but also drove opportunities for resolution. Water has been the reason for cooperation in the past and could possibly be the reason for a sustainable and robust peace settlement between the Israelis and Palestinians in the future. Although the crux of the issue is related to ending the occupation of Palestine and achieving sovereignty and stability, the Palestinian-Israeli conflict has been fueled by the issue of inequitable access to water. By the same token, I argue that addressing the issue of water allocation between Palestine and Israel will have immeasurable positive impacts on the overall geopolitical arena. As such, equitably resolving water resources allocation should be looked at as a steppingstone towards resolving these larger political issues, inducing land, the status of Jerusalem and the right of return that the two nations have been contending with for decades now. Water should be a fundamental ingredient in any future peace-making effort, as hard as it is to imagine given the current bleak prospects for peace and reconciliation. This calls for both the Palestinians and Israelis to divulge and clarify their intentions, positions, and interests pertaining to the establishment of a robust mechanism capable of producing and sustaining peace, whatever that might be.

The key to landing a deal that resonates with everyone is to find creative solutions rather than looking for pure compromises that each party can offer. The land for peace formula failed to offer a reasonable deal because it was based on the assumption that both sides must offer some compromises to achieve peace. The issue with that assumption was that each party has a starting point baseline that was very different from the other party, from which a point of departure was measured to discern the magnitude of each party's respective compromise. For the other party, however, that may not necessarily be considered a compromise but rather a given, because their respective baselines and points of departure were vastly different. As such, the two parties need to look for creativity rather than compromises in their search for solutions, alternatives and possible agreements. With a positive-sum approach, a lasting agreement that makes everyone a winner is what we are looking for here. That winning formula must be tied to international laws (namely UN resolutions 242, 338, and 194) which call for the complete withdrawal of the Israeli army from the territories it occupied in 1967 and providing a "just settlement" to the refugee problem, coupled with an acknowledgment that the current situation cannot be used as a baseline for negotiation. Namely the dismantlement of the illegal Israeli settlements in the occupied territories must not be viewed or presented as a compromise that Israel is offering to move negotiation towards a settlement. Far from being compromises or preconditions, these are expected legal obligations and prerequisites for Israel to abide by prior to entering any negotiations with the Palestinian side. These ground rules must be vividly outlined, communicated and understood by all parties entering into these negotiations.

It is worthwhile noting that the peace process was theoretically based on these principles. However, the parties failed to formulate a concrete agenda for their implementation and therefore failed to influence any change in reality. Although forcefully championed by the Palestinian negotiators, these principles were largely not adhered to. This hampered the negotiation and created insurmountable impasses that grew into confrontations and resentment whenever key issues, such as Israeli settlements for example, were brought to light. Applying and adhering to international law is mutually beneficial and means using UN resolutions 242 and 338 as the foundational principles for negotiation. It also means adhering to UN resolution 194 by either allowing the Palestinian refugees the option to return back to their original villages or receive compensation. Either way, this will be a great step not only for the Palestinians but also for Israel. This will boost Israel's demographic character and cultural diversity that will undoubtedly increase its legitimacy and image as a melting pot of multiple nationalities and heritages rather than the current exclusive identity. It will also provide a good-will measure that will increase the chances of acceptability and prevalence of a united vision for peace that we discussed in this book.

On the whole, the negotiation process takes generally three major phases. These include the pre-negotiation, negotiation, and post-negotiation, as shown in Fig. 1. Under the current situation that the conflict reached, there are numerous questions that need to be substantially addressed in future negotiations. These include, but are not limited to the following:

- How can the United States serve as a neutral third-party mediator considering its track record?
- Who can alternatively serve as a neutral third-party mediator in lieu of the United States ?

Fig. 1 Negotiation process phases

- Would a one-state solution or a two-state solution be the acceptable arrangement?
- If a two-state solution provides a more acceptable resolution, would a Palestinian state be established in the West Bank and Gaza Strip? How will these two separate geographic entities be connected and administrated? How will this territorial connection be governed, controlled, and secured?
- What's the status of Jerusalem? Who has sovereignty over the city? How will this control be allocated?
- What would happen to the current Palestinian refugees? What options will they be offered and how can this be implemented in light of the one-state versus two-state solution options?
- What will happen to the Israeli settlers in the West Bank? Will they be relocated inside Israel? How will these assets be managed and incorporated? What will happen to the wall?
- What's the status of the Arabs living inside of Israel? Where will they be located? Will they be given the same rights?
- How will the parties share international water resources, including surface water and groundwater?
- What's the solution to the Gaza water problem given its polluted and irreversibly damaged groundwater (which is the main source available to the two million people living in Gaza)?
- What needs to happen for Gaza to become "livable" again?
- In what ways can the Gaza economy be resurrected?
- How and who will pay for the reconstruction and rebuilding of the war-torn Gaza infrastructure?
- What would be the status of the Palestinian freedom of movement? Who will control border crossings and how will these be managed?
- What type of economic and political relationship would the parties have in the future?
- What is the mutual vision for the future of the region as a whole?
- How will these agreements be implemented and enforced? Who will ensure their enforcement and what enforcement mechanism will be used to accomplish implementation? Who will guarantee this and how?

- What would happen to the party that violates these agreements? What are the ramifications?
- How often will these agreements be revisited?

2.1 Pre-negotiation

During the pre-negotiation process, all affected parties including the independent representation of both Palestinians and Israelis in the process to avoid the adverse impacts that could result in the ex parte syndrome by excluding important stakeholders. Utilizing confidence building measures (CBMs) and evidence-based decisions are critical components of the process in this stage to ensure trust and objectivity respectively. The pre-negotiation phase provides a great platform for the parties to separate people from the problem, create alliances, establish priorities, promote a maniacal focus on a united vision for peace, and cement a cadence of accountability for a mutually desirable outcome. It also provides an opportunity to agree on the problem definition and negotiation agenda, as well as formulate and agree on a united vision for peace. Validating and testing assumptions about problems early and often is critical to minimize the cost of failure, which in this situation can be consequential as we have seen in the past. At this stage parties need agreement on an agenda that will give them enough airlift capacity to produce focus and drive a plan of execution. A number of key issues must be collaboratively explored and amicably addressed, including the one-state versus two-state solution, exposing loyalties and finding a common purpose, fostering adaptive learning while neutralizing emotions, and exploration of options for mutual gains.

First: One-State Versus Two-State Solution

The discussions about the one-state versus the two-state solution drove unnecessary yet damaging tumult about the negotiation process. None of these solutions should be discussed in separation of the current situation on the ground and the formation process of any potential agreement. These discussions must all be framed in the context of negotiation and comprehensively evaluated to identify their viability, acceptability, practicality and tradeoffs. This has yet to happen in any of these discussions on the topic, and should be addressed as part of the pre-negotiation process. These and other options must be objectively considered and assessed in light of agreed upon evaluation criteria that consider all four factors identified above.

First, the viability of the solutions must consider the overall context within which the proposed solution is applied and the sustained future success of its implementation. This should be tightly coupled with the financial and economic elements of how to build healthy economy capable of leveraging the country's assets and resources to create independence and autonomy. Under the one-state solution, it is unclear how these important aspects related to its viability will be configured, managed, and administrated. Second, the viability and success of any future negotiated settlement is primarily hinged on its acceptability by both sides. For many people who are involved in the conflict, the thought of a one-state solution is either unimaginable or premature to even consider at this stage. The current political and ideological landscape is so divisive that the acceptability of a one-state solution poses a particularly troubling aspect that has yet to be tested in reality.

Third, the practicality of the proposed solutions must also be evaluated in light of their respective tradeoffs. Currently, the two-state solution (which is s US-based approach) is logistically impossible without a total dismantlement of illegal settlements and putting an end to the occupation before entering into any further negotiations. Otherwise, the talks will not yield any substantive results for either side and will inevitably produce the negotiation paralysis situation that has characterized the peace talks in the past twenty years or so. On the other end of the spectrum, the one-state solution, which is getting more traction these days (as a result of the logistical difficulty and practicality associated with implementing the two-state solution induced by settlement buildings in the occupied territories) will neither offer the Palestinians their self-determination they seek, nor will it preserve Israel's identity as a Jewish state. For Israel, the acceptance of the Palestinian right of return and the mere presence of a Palestinian majority would not only amplify the de jure apartheid that is enshrined and reinforced on the ground, but it would also upend the very identity of a Jewish majority demographic composition. As such, the one-state solution will, ipso facto, force Israel to make a choice between being an all-inclusive democratic state or a Jewish state, neither is attainable considering the incontestable evidence pertaining to Israel's track record with its Arab citizens who are struggling as a result of its systematic discrimination.

The kerfuffle surrounding the one-state solution primarily emanates from the fact that the solution assumed the inevitability of the current unjust situation, namely the illegal settlements in the West Bank, which is taken for granted as an unavoidable, irremediable reality. The focus of the discussion should be shifted towards ways to compel Israel to freeze and dismantle its illegal settlements in the occupied territories in accordance with the international law, instead of treating the settlements as a reality in need of legitimization. Reflecting the current settler-driven agenda, Israel wants to "negotiate" with the Palestinians without any pre-conditions. The issue with that request is the fact that this has been precisely the problem with all previous negotiations. Dismantling the illegal settlements in the West Bank and East Jerusalem and ending the occupation must not be viewed as pre-conditions—they are legal obligations that Israel needs to abide by before the resurrection of any future negotiations. Otherwise, we will end up with this same stalemate and more practical issues related to the plausibility of implementation. The persistent Israeli refusal to halt its settlements, end its occupation of the Palestinians and abide by the international law continues to constitute an insuperable obstacle to implementing the two-state solution and overall peace, and its removal is key to achieving peace.

While valuable and necessary to consider, the one-state solution discussion so far failed to offer remedies for the hard reality on the ground, which is further complicated by the building of these settlements and the separation wall, by avoiding it all together. Another troubling aspect about the one-state solution discussions is that they all lack substance and have not considered its practicality and acceptability by both sides. The fact that these two sides have been fighting for the past decade or so and the ongoing fundamental disagreement even regarding the two-state solution are a clear indicator of the likely failure of the one-state solution to gain acceptance from either side, although being seriously considered by some scholars and practitioners. It is also unclear at this point, how the parties will govern themselves under the one-state solution, namely the institutional design and governance of the state and the question of representation. The issues of governance and representation are thorny and troubling in Israel today, which are surrounded by allegations of systematic discrimination towards the Palestinians living inside Israel. This alone makes the one-state solution discussions untenably divorced from the reality of the discourse at present time.

Along this vein and expressing his country's worries about the ongoing takeover of the West Bank for the exclusive use of Israeli settlements, the U.S. Secretary of State John Kerry stressed their detrimental impact to the viability of the two-state solution which he sees to be also a threat to the overall prospects of real peace in the future. His appeal to both sides was to resurrect the implementation of the two-state solution which many consider to be near death. He further laid out a comprehensive vision in what appears to be a last ditch push for peace in the Middle East. This included several principals to ensure that the two-state solution is not dead in the water. These principles include[1]:

1. Securing recognized international borders between Israel and a viable and contiguous Palestine based on the 1967 lines guided by the UN Resolution 242;
2. Creating two states for two peoples with mutual recognition and full equal rights for all their citizens to fulfill the vision of UN General Assembly Resolution 181;
3. Finding a just, agreed, fair and realistic solution to the Palestinian refugee issue, with the support from the international community including compensation and assistance in finding permanent homes;
4. Providing a resolution for the status of Jerusalem as the internationally recognized capital of the two states, and protect and assure freedom of access to the holy sites;
5. Ensuring Israel's security and bringing a full end to the occupation; and
6. Ending the conflict and all outstanding claims in order to enable normalized relations and regional security for all as envisaged by the Arab Peace Initiative.

Second: Exposing Loyalties and Finding a Common Purpose

Every party comes to the negotiation table with a set of declared and hidden loyalties, motives, and interests. Exposing and fully understanding these elements is crucial in aligning everyone behind a common purpose. This simply takes time and disciplined negotiation, and depending on the level of trust, fully exposing them may not be possible. However, unearthing these fundamental elements helps the parties to

[1]For the complete transcript of the U.S. Secretary of State John Kerry's speech on Dec 28, 2016, in which he presented six points for future peace deal, outlining the U.S. position towards resolving the Palestinian-Israeli conflict, refer to Appendix B.

understand each other's positions, interests and priorities in order to reach a common purpose for the negotiation. This common purpose is pivotal in orienting the parties and providing a winnable agenda that all parties can endorse and implement. Furthermore, understanding and neutralizing conflicting and competing positions can provide a platform to nurture an interest-based approach to addressing all parties' needs with an eye on practical ways to achieve this common purpose throughout the entire negotiation and treaty implementation lifecycle.

Third: Fostering Adaptive Learning While Neutralizing Emotions

Fruitful and impactful negotiation calls for a process of adaptive learning. While this adaptive learning process is aimed at learning about the process, the parties, their interests and loyalties, as well as about alternatives and creative solutions, it can be triggered by "disturbing the equilibrium" to move things beyond the status quo. However, this disturbance cannot exceed a certain level of emotions that would threaten the overall stability of the process and the adaptive learning range. If exceeded, this level of emotions can be devastating to the learning process and its outcome, as parties would tend to react emotionally and aggressively. This emotional range must be avoided to keep things within the acceptable zone of adaptive learning for some time, in order to allow the parties to gain progress towards achieving their common purpose.

2.2 *Negotiation: A Framework for a Transformative Approach to Conflict Resolution*

The examination of the Palestinian-Israeli peace process and previous negotiation attempts that led to the perennial deadlock is a testament to the complete and utter failure of these diplomatic efforts that were conducted under the aegis of the U.S. Because of the inadequacy of these previous conflict resolution attempts between Palestine and Israel, many issues still dominate the diplomatic and geopolitical arena related to the grinding military occupation that the Palestinian communities are forced to live under, illegal settlement expansion, and inequitable distribution of land and resources, to name a few. The current situation that the conflict reached serves as a ubiquitous invocation and the impetus for settling this dispute once and for all with a comprehensive and multi-scalar approach. A renewed effort needs to recognize the feebleness of the previous process, draw lessons and parallels, and promote a holistic approach to clear the fog of obfuscation and resentment that surrounds the Palestinian-Israeli relationship. As such, a transformative approach to negotiation is warranted to enable the parties to find new sources of value that they can collaboratively chase and capture. It is also crucial to underscore the importance of reaching a just and lasting agreement that is, and pari passu with this, enforceable and legally binding so neither the Palestinians nor the Israelis can undermine it or take a position that can be viewed as ultra vires the terms of the settlement. Transformation does not mean perfection. Here we are not looking for perfection but rather measurable, quantifiable and demonstrable progress. We need not confuse optimism with the atrocious reality that is demonstrated on the ground as a result of the ongoing occupation and the evident failure of the so called peace process. Transformation that I seek here is concerned with improving the situation for both nations—for Palestine to achieve statehood and for Israel to achieve security. It is the ability of these two nations to trade a past engulfed with hostility and restatement with a more hopeful and promising future that brings reconciliation, justice and peace for all.

The complex nature of this intractable conflict including land and water issues provides ample justification for the need to develop a new agenda for a transformative, interdisciplinary, and pragmatic approach to conflict resolution. This agenda must recognize the uniqueness of this conflict and the challenges surrounding the key issues in dispute, while at the same time providing guidance for practical and fruitful negotiation approaches to foster equitable agreement formation and decision-making processes. This necessitates delineating practical boundaries for a transformative approach that moves the Palestinian-Israeli conflict from its current hostile, non-cooperative state to a more amiable and cooperative state. The real value of defining these new boundaries to chart the course to a more successful negotiation practice lies in the ability to engender recognizable improvement in the quality of the agreements and ultimately achieve equitable processes and outcomes.

It is commonly believed that water is the source of conflicts, and this conflict is no exception. However, this is not entirely true considering that water is an obstacle to peace rather than the main source of the dispute. By the same token, I argue that removing this obstacle and resolving the issue of water allocation can effectively serve as a catalyst for a just and lasting peace between the Palestinians and Israelis. Because of its boundary-crossing nature, water can present a set of complex challenges related to equitable water distribution to adequately and sustainably respond to a rapid population growth pattern with its exponentially increasing water demand levels. Water is tied to many dimensions such as population growth, economic development, agricultural production, and even national security, to name a few. The multidimensional aspects of the complex, uncertain, and uncontrollable political, natural, economic and social realities of water suggest the need to develop a new framework that encapsulates the merits and imperatives of a multifaceted approach as well (Islam and Susskind 2013). Issues related to water demand, supply and equitable distribution and allocation are also in flux, and as such, our intervention needs to recognize the changeable nature of these complex issues and have the necessary agility to respond to these changes. Akin to selling umbrellas in the rain, the application of equitable water-sharing to disputed water resources in the Middle East is of paramount importance in order to achieve hydropolitical stability and advance an acceptable united vision for peace. Equitable sharing of water resources, therefore, tends to induce cooperation among nations involved in high tension political agons, which sets the stage for peaceable relationships among them. Equity in water resources allocation and the ongoing pursuit of sustainable peace in the Middles East are inextricably linked.

It all starts with greater knowledge of what went wrong, awareness of the opportunities, and evaluation of adequate and cross-disciplinary options to resolve the dispute. Reaching a negotiated settlement that is just, lasting and robust necessitates that the parties engage in an ongoing dialog in order to produce a treaty formation process capable of building the right agreement, building it right and building it forward with an outlook at fostering and maintaining successful implementation. The way this must be approached is by meeting three essential criteria including the need to abide by international laws, which must be perceived as a prerequisite rather than a pre-condition, utilizing water as a catalyst for peace by allocating it equitably between the Israelis and Palestinians, and formulating a united vision for peace that both sides can rally their respective nations behind. For that to happen, a pragmatic framework for transformative conflict resolution is proposed in Fig. 2 that encapsulates a trilogy of key elements including rules of engagement, mechanisms of engagement and the involvement of a neutral third-party. This framework provides a list of guiding principles inspired by professional and personal experiences. As explained earlier in this book, this approach is not meant to be a solution to the conflict. Nor does it represent the best framework to be applied here. This framework only provides a toolkit for a roadmap for peace, and as such, its success primarily hinges on the parties' willingness and ability to execute and implement it in a meaningful and disciplined way to achieve mutual gains. It is pivotal to reinforce what has been already discussed in this book that unless such a practical framework is utilized in conjunction with a set of prerequisites outlined earlier, the fate of any future negotiations between the Palestinians and Israelis would not be any better than what twenty five years of futile negotiations has yielded.

Crucial in this context is the objective of this approach which is not only confined to alleviating water stress, but it also includes a myriad of strategic, political, and economic gains towards equitably resolving the conflict. In essence, this approach aims to *equally* increase the net benefit from water allocation, in all its forms, for *all* stakeholders involved in this dispute as a way of bringing understanding and consensus over other high-politics issues. The assumption made here is that equitably resolving water allocation will have positive impact on resolving the land allocation and by extension the issues of sovereignty and national security, namely that water is tightly linked to all of these issues. The extent to which each country benefits from a water allocation scheme is primarily determined by the current water needs met measured by water stress alleviated for both sides. In an ideal world, maximizing water stress reduction for each country means that the reduction in water stress should be proportional to water needs. Another key goal relevant to the scope of this approach is to *equally* decrease harm, in *all* its forms, for *all* parties. As an essential prerequisite, a mature adaptive application of equitable water negotiation parameters should incorporate an examination of international water negotiation trends in the past, the relationship between political and economic forces and cultural and social structures, and ways in which power relationships shape political realities

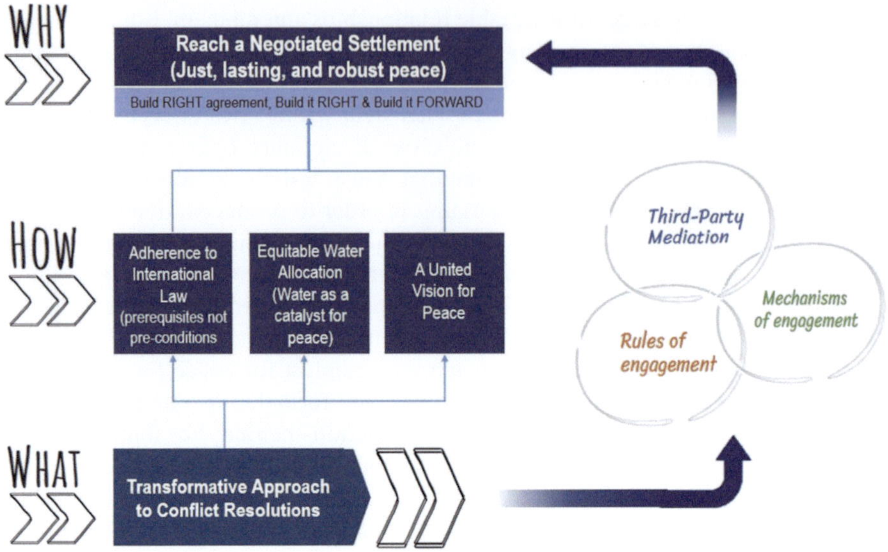

Fig. 2 Pragmatic framework for transformative conflict resolution

and decision-making regarding water allocation and distribution (Abukhater 2013). While discourse can provide part of the explanation, a further step is to foster better examination of the structures of power (or hydro-hegemony) that not only shape discourse, but in many cases generate it. With this in mind, we must be visionary, creative and pragmatic, attentive to both process and discourse, and cognizant of the seminal role of hydro-political structures and dynamics of this particular conflict. The fact that this transformative approach to conflict resolution is equally concerned with short and long-term consequences of water allocation to promote integrated water management lends it credibility and merits to be influential in transforming the rather stagnant and hostile relationship into a more cooperative and fruitful one.

By utilizing equitable water-sharing criteria, this transformative approach includes three key components that, together, can produce conducive negotiation and decision environments in order to achieve optimal outcomes. These include the jurisprudential framework of water law and water institutions amalgamated to adaptive learning and consensus building, equitable water-sharing criteria, and third-party mediation. This comprehensive nature of this approach makes it exceptionally functional and relevant to the Palestinian-Israeli conflict. As depicted in Fig. 3, this framework has three essential elements that when meaningfully applied will yield demonstrable results. First, applying principles of the international water law provides process guidance, or "*rules of engagement*" to level the playing field and eliminate legislative issues and deadlocks. Second, adaptive learning and application of equitable water-sharing criteria provides institutional resilience and governance, or *"mechanisms of engagement,"* in order to eliminate hydro-hostility and promote hydro-stability. Third, utilizing effective ADR and the involvement of a neutral third-party mediator are essential elements for attaining win-win cooperative processes in order to yield acceptable resolutions for both nations. As explained in my previous book, *Water as a Catalyst for Peace*, there are a number of important procedural equity aspects coupled with institutional principles that, together, play a seminal role in providing a conducive atmosphere for effective negotiation with equal footing, and the ability to shape outcomes with the flexibility to accommodate and respond to different political, social, economic, natural, and spatial contexts, on one hand, and the aspiration to promote legitimate public feedback in pursuit of the common good, on the other hand.

First: Rules of Engagement

Pacta sunt servanda, the basic goal of the international law is to achieve applicability of its principles so they are kept and adhered to by the parties. The first cardinal step in future negotiation between the Palestinians and Israelis is to use a common set of agreed upon ground rules to govern their interaction and to eliminate any chances for conflict of interests during or after negotiation. This step was not only missing but it was deliberately overlooked in the previous negotiation attempts that the Palestinians and Israelis undertook to bring their century-long conflict to an end, which in turn caused difficulties and disagreement, ultimately leading to their demise. Eager to adopt the principles of international law, the Palestinian officials prematurely tried to refer to these laws as relevant international legal norms to be applied to the bilateral negotiation process in which they were engaged without first securing their Israeli counterparts' consensus on their applicability. This was amalgamated by the resounding skepticism among Israeli negotiators about the applicability and usefulness of international law to govern their interaction, as they perceived it irrelevant on the basis that it did not offer a clear template for negotiation. This was detrimental not only to their interaction but also, and more fundamentally to the outcome of their negotiations and the enforcement of agreements thereafter.

Although the United Nations General Assembly passed a number of resolutions that delineated "principles for the achievement of comprehensive peace," these principles were for the most part rejected by Israel. These principles include the following:

(a) The withdrawal of Israel from the Palestinian territory occupied since 1967, including the Gaza Strip, West Bank, and Jerusalem;
(b) Arrangements for security assurance of all states in the region within secure and internationally recognized boundaries, including those mentioned in resolution 181 of 1947;
(c) Rectifying the problem of the Palestinian refugees in keeping with the General Assembly resolution 194 of 1948; and

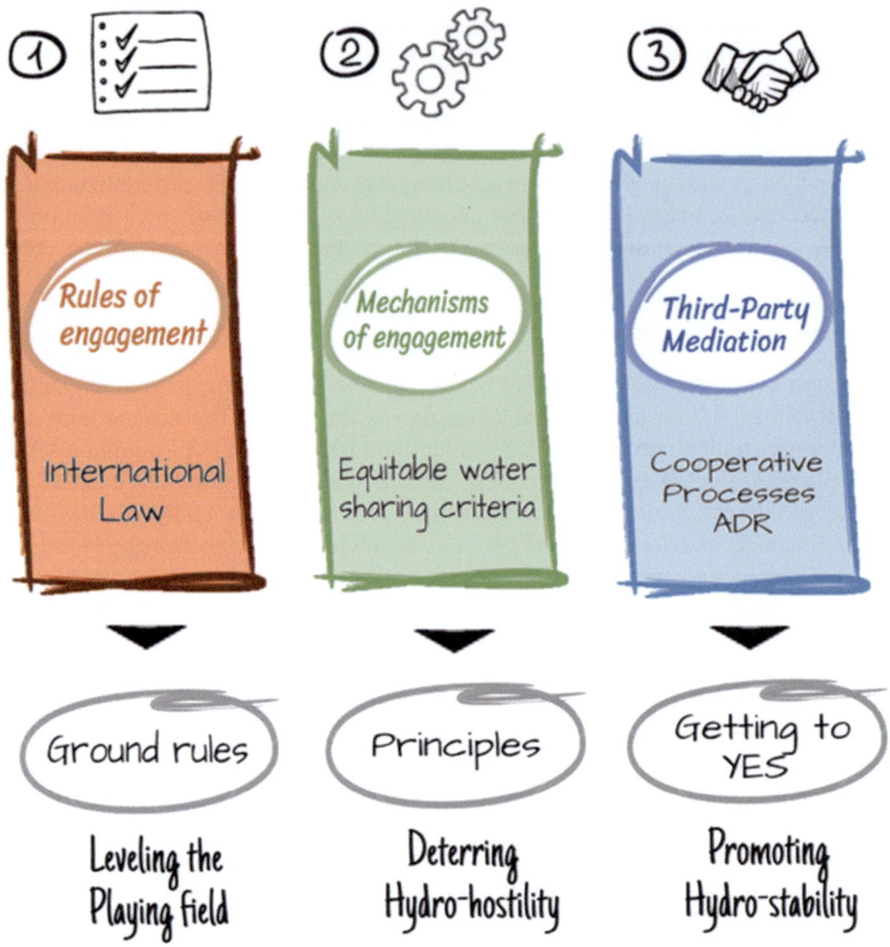

Fig. 3 Components of a transformative approach to conflict resolution

(d) Dismantling the Israeli settlements in the territories occupied since 1967 and facilitating freedom of access to holy places and religious sites.

From the onset of their diplomatic effort, the two parties expressed sharply differing views regarding the relevance and applicability of these international legal norms, and by extension their respective rights and obligations under these laws (Dajani 2007). Disagreements arose over the meaning and interpretation of international law, as it was deliberately used to justify and legitimize opposing positions. However, the existence of differences between negotiating parties regarding the interpretation of international law does not necessarily indicate that the law itself is indeterminate (Dajani 2007). It simply means that the parties have different interpretations and in need of a third party who can interpret and enforce the law. The failure of previous negotiations to achieve consensus regarding the application of international law to the Palestinian-Israeli context made it increasingly difficult to create a frame of reference that the parties can defer to in case of inevitable disagreements during these negotiations.

International law can shape both the process and outcome of peace negotiations not only as a result of its enforceability but also as a result of the guiding principles and structure that it may offer the parties to formulate a negotiated outcome irrespective of its enforcement (Dajani 2007). In this sense, legal rules contribute to zone definition and each party's reservation point which is defined by their "best alternative to a negotiated agreement" (BATNA). This legal influence is referred to by Mnookin and Komhauser as the "shadow of the law," which akin to social norms and customary practices, can determine what is being considered or omitted from an agreement (Mnookin and Kornhauser 1979). Although the applicability of international law is inconclusive, it at least provides general rules to allow room for flexibility and adaptability so that states can more easily find solutions that can work for them in various cases. As such, legal norms may provide standards for resources allocation, as Joaquin Tacsan astutely explains, "though parties to international disputes have certainly been known to manipulate international law in decidedly non-legal ways, such parties often enough choose to bargain on legal grounds because international law permits less of a range of manipulative calculation than other types of bargaining" (Tacsan 1991).

International law may exert influence through its perceived value by the parties, regardless of its enforcement, by providing objective standards for resolving disputed issues such as water allocation. According to Dajani, 2007, "Although the recognition of legal rights should inform the framework for negotiations, it neither obviates the need to explain the legitimacy and justness of the rights invoked nor should it foreclose discussion of equitable interests that may justify a departure from the remedy prescribed by law." In this spirit, international law is not only applicable but it is also very useful to facilitate equitable negotiations and reconcile conflicting interests. As such, relying on international law is a prerequisite to safeguard the process of negotiation and more importantly, provide the legitimate framework for negotiation and necessary protection for the weaker party throughout the negotiation lifecycle.

In the context of negotiation over water sharing, effective future talks should be based on the adaptive application of the guiding principles of the international water law as an overarching jurisprudential framework. This step is essential in providing the necessary "rules of engagement" to level the playing field and neutralize or eliminate the impact of power-structure imbalance. It is often the case that most powerful states reject the applicability of international law, as we have seen in the Palestinian-Israeli negotiations. Applying international water law can provide a useful instrument to effectively mitigate the impact of hydro-hegemony by promoting transparency, accountability and predictability of behavioral tendencies of powerful states. To that end, Woodhouse, 2008, argues that relinquishing international law altogether without exploring other mechanisms to govern international negotiation over transboundary water tends to not only undermine the regional water security, but it may also lead to damaging the political relations among riparian states.

Encapsulating economic, hydrologic, and legal criteria, international water law, namely the Helsinki Rules of 1966 and the 1997 United Nations Convention on the Non-Navigational Uses of International Watercourses, delineated a number of useful and practical factors for equitable allocation of transboundary water that can potentially be applied and utilized as a template against which various water allocation alternatives can be assessed. Concerned with enhancing "reasonable and equitable share in the beneficial uses of the waters of an international drainage basin," these laws outlined nine factors as key parameters for equitable water allocation and utilization. These include population, hydrology, climate, historical and existing uses, socioeconomic needs, dependent population, comparative costs of alternative means, availability of other resources, and the degree of appreciable harm or damage caused by denial of water rights (International Law Association 1967). Operationalizing these nine factors of equitable water allocation means applying them as a meaningful legal foundation for water negotiation between the Palestinians and Israelis. These factors, referred to as drivers of outcome equity, as shown in Fig. 4, can potentially provide a "template" for estimating judicially cognizable water needs for negotiation that was missing in previous negotiations and is essential to advance negotiation to more plausible and implementable resolutions. By anchoring the negotiation process based on this international law-driven template for equitable water negotiation, both sides can consider and capture these guiding elements to allow flexibility as well as consistency in equal measure. This reliance of principles of international laws is necessary yet not sufficient in and of itself to overcome the competitive arousal trap—that is the desire to win at any cost (Malhotra and Bazerman 2008). As such, this must be accompanied by goodwill, a neutral third-party mediation, and a maniacal commitment to a united vision for peace as an integral component of efforts to build popular support for peace.

Second: Mechanisms of Engagement—*A Comprehensive Approach to Conflict Resolution*

Mechanisms of engagement incorporate a practical set of adaptive and operational elements that personify instrumental equity parameters pivotal to the success of future negotiation. These key elements are compiled based on a careful examination of many treaties coupled with awareness of the pitfalls of the failed previous negotiations between Israel and Palestine that led to the current political and diplomatic impasse. The lessons and parallels learned from these cases serve as substantive evidence that informed this transformative framework. The ability of the Palestinians and Israelis to lead successful negotiation capable of sustainably and effectively resolving their conflict with potential agreements that are desirable, viable, and equitable, primarily hinges on their ability to utilize an integrative, comprehensive approach to conflict resolution grounded in a number of crucial components. These components, illustrated in Fig. 5, encompass context-sensitive equity and adaptive

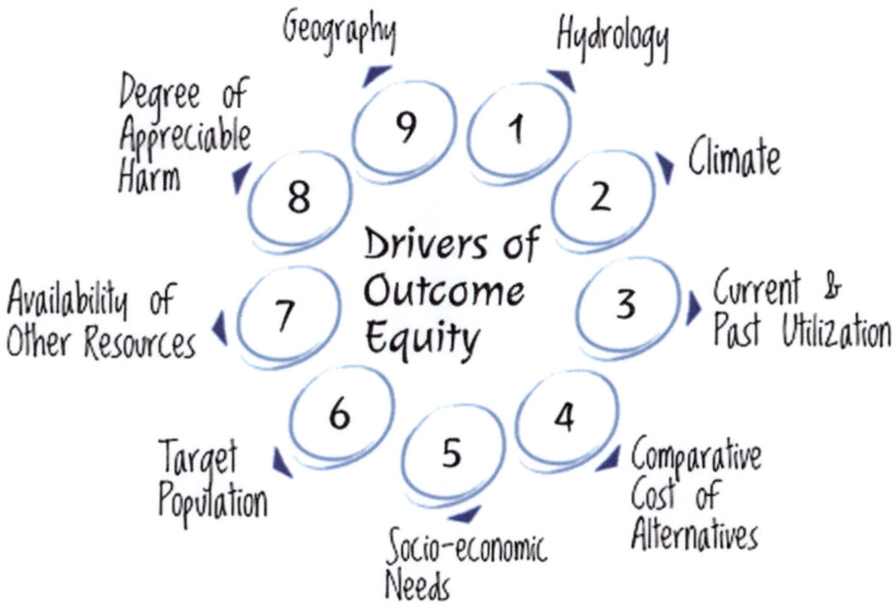

Fig. 4 International law-driven template for equitable water negotiation

learning and management, modern and indigenous approaches to conflict resolution, and goodwill/faith-based confidence building measures (CBMs), on one hand, and incorporating an incremental negotiation process, sectoral prioritization of various uses and users, linkage to environmental and non-environmental issues, *sharing* not *dividing*, needs-based water allocation, and satiety-based hydro-diplomacy, on the other. As I describe these components of this comprehensive approach in the following section, I will attempt to tie them to specific issues pertaining to the current Palestinian-Israeli conflict in order to shed light on how, in practice, these components can and should be used and implemented in the context of future peace negotiations. The goal is not to provide a cut and dried solution or a magic wand that somehow will predictably resolve the conflict, but rather a set of guiding principles that, when applied carefully and thoughtfully, can potentially enable the two nations to mutually identify, create, augment, capture and communicate value.

1. **Incremental negotiation process**

Water disputes are complex in nature. What makes them even more complex is the dynamic nature of these disputes that usually evolve over time creating new circumstances and realities and necessitating new strategies and ways for intervention as time goes by. Reflecting this dynamic nature, both the principle and practice of water allocation and dispute resolution evolve along two spectra—the number of key stakeholders, and the number, nature, and magnitude of issues involved (Shmueli 1999). This dynamic nature of water allocation and the evolution of associated disputes

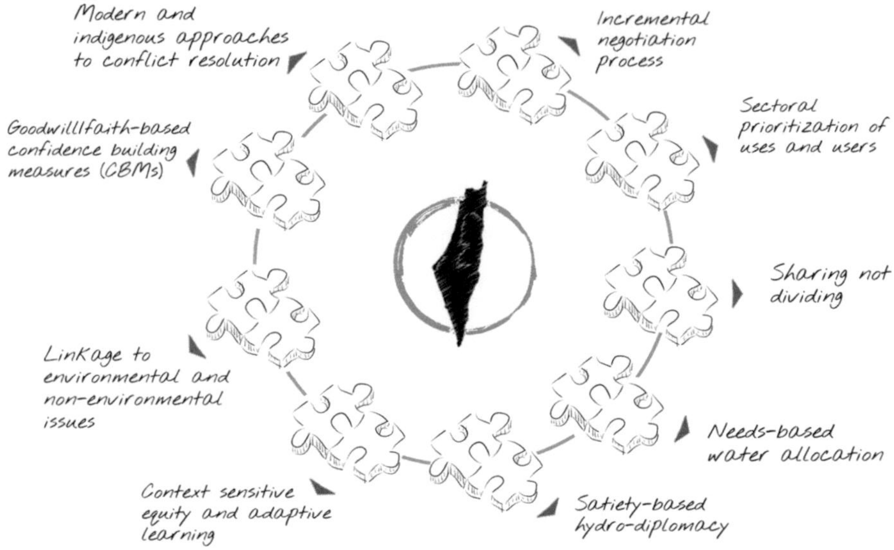

Fig. 5 Mechanisms of engagement—comprehensive approach to conflict resolution

and techniques to resolve them are reflective of the complex nature of water itself as essential for human survival and economic development. Thus, negotiation over water (as a bridge to peace) requires an incremental approach to mirror and respond to this dynamic nature of water disputes. This is particularly important during the initial stages of negotiations as disputing parties often hold vastly different interpretations of the issues at hand that need to be addressed and resolved. This divergence of interpretation can cause unwarranted tension as parties may seem unwilling to cooperate with each other, or worse yet, could lead them to discount negotiation as a viable option altogether. Securing agreement on these key issues is critical and can take time. The parties' ability to reach an agreement regarding common issues of concern at the beginning of the negotiation process is crucial to its success. This can be accomplished through an incremental negotiation process that realizes the divergence of parties' interpretation, the evolving nature of these very issues, and the need to collaboratively achieve consensus to ensure negotiating parties are fully aligned and issues are adequately identified.

With this in mind, incrementalism is a fundamental practical concept that needs to be incorporated and implemented from the onset of negotiation. Since creating and adopting new agreements to govern the allocation of disputed water resources cannot be accomplished in a vacuum, attempting to address all issues with all parties at once can create diplomatic deadlocks that can jeopardize not only the negotiation process but potentially the relationship between the parties involved. Conducting piece-meal negotiation means involving different parties while addressing different issues over the course of many years and many rounds throughout the entire negotiation lifecycle.

In the context of the Palestinian-Israeli negotiation, this concept has been grossly misunderstood and misled the entire process and its failed outcome. The examination of the Palestinian-Israeli negotiation process provided irrefutable evidence of the failure to apply this incremental process adequately. The parties, primarily steered by Israel's own interests, artificially focused on trivial issues and deliberately ignored the core issues at hand, such as settlements, Jerusalem, water resources, and the right of return for the Palestinian refugees. Instead, under this false implementation of the concept of incremental negotiation, the issue, as we discussed previously, became concerned with the process itself rather than achieving tangible peace. As a result, this incremental negotiation concept was used to justify negotiation without real outcomes. Everything that came out of that process has been in essence of the interim sort, including the formation of the Palestinian Authority and the alleged sovereignty. This approach was not adequately applied in this case and what was thought to be an incremental approach to negotiation demonstrated serious shortcomings that amounted to unequivocal failure. The interim situation that was achieved was not used as an opportunity to build trust and chart the course to the final status negotiation. Instead, this interim period was used by both sides as a platform to maximize their respective bargaining position and therefore their endgame, without a clear line of sight into the finish line of where this whole process was heading. As a result of this, Israel was not willing to give up settlement construction, and Palestinians unwillingly acquiesced to crack down on armed Palestinian groups. Hatred and grudge bred more hated and grudge reactions from the other side, leading to the ultimate demise of these negotiations and the so called peace process altogether (Agha and Malley 2002).

There are pivotal lessons that can be learned here. In the spirit of the understanding that who begins with too much accomplishes little, it is important for the Palestinians and Israelis to relay the value of incorporating incremental negotiation process and an agenda to translate a untied vision for peace into reality by incrementally leading to resolving the conflict. Revolving the discussion around a problem not a solution is critical. A problem must be at the center of every discussion, rather than solutions. As such, the early stages of negotiation must be a problem-oriented negotiation instead of jumping to solutions to problems that may or may not be real. We need not reverse-engineer the solution to match a problem that may not exist. What problem did creating the Palestinian Authority solve? We must address the real problems, but first get agreement on them. Falling in love with a problem means focusing on the problems we solve not the solutions we create, which is the first step in this incremental approach.

Ameliorating conflicts of significant magnitude, such as the Palestinian-Israeli conflict, calls for a systematic process of incremental negotiation, by utilizing both initial and final agreements, reflecting the dynamic nature of water disputes. This incremental negotiation process must be grounded in a prioritized set of important issues that both sides agree on and that is capable of charting a roadmap that (as a whole) ultimately and *incrementally* can lead to resolving the conflict in its entirety. Taking small but measured steps towards accomplishing key goals is key to success. This concept of incrementalism, if applied correctly in negotiation, is greatly beneficial for many reasons. First, carving out smaller topics helps in simplifying the negotiation agenda and topics to be addressed. This simplification of topics and agenda that are easier to negotiate and resolve helps the parties to accomplish quick wins and address the low-hanging fruit before tackling more complex and controversial issues while building trust and credibility in the process and each other. This is because incrementalism helps in facilitating and building relationship based on the gradual accumulation of successes on small but key issues and milestones that map out to a larger and more impactful goal, bounded by a common understanding of the problem and a united vision for peace. Agreeing on basic issues and building trust and credibility is of great importance in the context of the Palestinian-Israeli conflict considering the current mistrust, its history and emotions involved.

Initial agreements are often formulated in order to respond to more urgent and immediate needs, while comprehensive agreements are commonly used to address long-term issues and overarching resolutions. In this regard, resolving a specific issue of concern, as minute and insignificant it might seem, can undoubtedly provide the parties a solid foundation and springboard for resolving more intractable matters. This is because gradually building on established agreements (such as bilateral agreements) can in time lead to reaching agreement on comprehensive ones (such as multilateral agreements). Equitably allocating water between Israel and Palestine can serve as this springboard on which the parties can build trust and credibility that has been missing in previous negotiations. This way, they can achieve basin-wide multilateral agreements in regards to water, as a relatively low-politics issue, which can serve as a way to foster the quality and sustainability of their diplomatic relations, on one hand, and provide a solid point of reference to resolving their most pressing issues, on the other. This steady incremental process can slowly but surely move the parties towards a more just, stable and permanent resolution. It is important to note that it is ultimately the parties' decision, with the help of an impartial mediator, to identify the appropriate agenda for their negotiation, what issues to be discussed first, and how resolving these core issues can cumulatively lead to the resolution of their dispute.

2. **Sectoral prioritization of uses and users**

The increasing demand on water resources drives the need to prioritize water allocation for different uses and users. Typically, downstream and historic uses are given priority. However, there are a plethora of competing realities and concerns that need to be amply addressed and delicately balanced in negotiation over water allocation, such as economic, social, ethical, and environmental dimensions. These are critical components that greatly influence the extent to which water is distributed equitably, efficiently, and sustainably. Generating and evaluating different water allocation scenarios are warranted to allow the two sides to augment the pool of alternatives available to them as well as their overall gain, instead of restricting their decisions to options currently available to them. The two elements that are central to approaching the issue of water allocation in the context of the Palestinian-Israeli conflict are concerned with prioritizing water allocation among different uses, on one hand, and striking a balance between productive human and environmental water needs, on the other.

First, adopting an approach that fosters water-sharing equity necessitates sound prioritization of water resources allocation between different sectors. In this arid region, the main competing sectors for water demand are domestic and agricultural uses. Decisions regarding allocating water for each of these sectors must take into account the magnitude of the current crisis in Gaza in particular and the larger Palestinian territories in general. Higher priority must be given to the domestic needs for those underserved communities to the extent needed to meet their current and future water needs. The key issue remains inexplicably interconnected to the question of necessity and priority. This issue must be looked at with a great deal of emphasis in order to achieve equitable allocation in comparison to the per capita water available for the Israeli domestic use. As such, sound water management calls for a prioritization of water allocation between the Israelis and Palestinians to ensure that highest priority is given to domestic purposes, followed by irrigation and industrial uses. In this context, securing uninterrupted access of adequate water, in terms of quantity and quality, for domestic uses at comparatively low prices must be given higher priority over other uses. Agricultural or industrial water uses cannot be given a priority irrespective of how profitable they are to the economic development of the region when we are confronted with severe water shortages on the Palestinian side that suffers from serious lack of domestic water to cater to their very basic needs. Economic benefits of agricultural water use in terms of food production and security, as important as they might be, cannot outweigh the domestic water needs namely when the latter is not adequately met, as is the case in the Palestinian communities. Future negotiation between the two nations must recognize the importance and urgency of meeting the pending Palestinian communities' domestic water needs as a first step, and then and only then will they be able to consider equitable criteria for prioritizing other uses as they see fit.

The second key issue that needs to be addressed is related to water allocation between productive and environmental uses, given the importance of water to both. Fundamentally, water is the sustainer of both human life and the health and well-being of ecosystems. This means that decisions regarding water allocations must also take into account balancing human uses and environmental concerns in a way that ensures improving and maintaining water quality and access, while fostering economic viability for both communities. We have seen many examples, such as the 1994 Israel-Jordan Treaty, where the negotiation process led by technocrats failed to consider environmental factors. These technocratic decisions often neglect the very simple fact that nature still has to compete for its fair share of water to maintain healthy and functional ecosystems. Economic and ecological needs often clash when people attempt to engineer and alter nature. The results are often disastrous and destructive to both nature and man. Not only did these negotiations contribute to the ongoing deterioration of water quality and the overall health of the aquatic ecosystem, they further propagated an unfavorable perception of their outcomes as being environmentally unfriendly, ultimately leading to their utter failure. While economic uses are important, negotiators must recognize (beyond lip service) that protecting the environment must not be an afterthought and should be given a priority in any decisions regarding water allocation schemes in order to maintain the integrity of

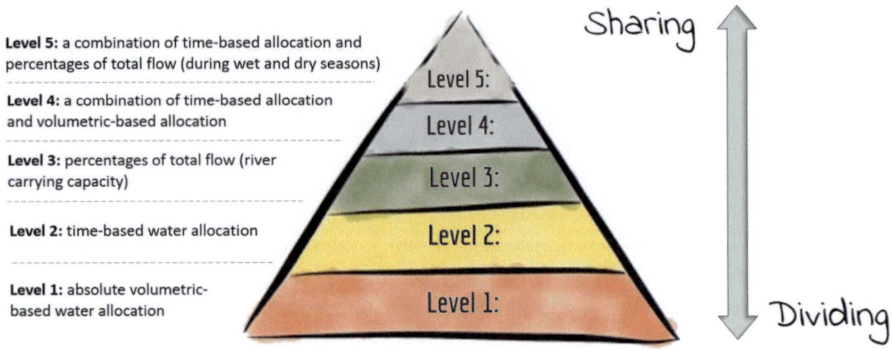

Fig. 6 The transboundary water allocation mechanism hierarchy

the environment through sustainable management of water resources. Illustrating the importance of water for the environment, Sandra Postel, 1997, in her book, *Last Oasis*, astutely contends that "we have been quick to assume rights to use water but slow to recognize obligations to preserve and protect it…. In short, we need a water ethic - a guide to right conduct in the face of complex decisions about natural systems we do not and cannot fully understand."

3. **Sharing not dividing**

Depending on the level of negotiation maturity, water can be either be *shared* or *divided*. There is a fundamental difference between *sharing* and *dividing* water. *Sharing* water is a desirable water allocation mechanism for many reasons. *Sharing* tends to foster a spirit of cooperation as it evenly distributes the benefit and risk between the two parties. It also compels the parties to look for ways to augment the pie and the spectrum of options available to them, ultimately leading to win-win solutions. One such example is to consider supply augmentation mechanisms to "increase the pie," rather than the consumptive division of available resources, which can inevitably lead to resentment fueled by the tragedy of the commons situation and depletion of these resources. *Sharing* means that both parties have the responsibility to work together to find ways to increase water supply and reduce demand. In contrast, *dividing* water resources tends to fuel competition and hostility and often overburdens one party with the risk of droughts or other financial responsibilities. As such, this approach is less desirable and must be avoided in future negotiations. In my previous book, *Water as a Catalyst for Peace*, I introduced five levels of water allocation mechanisms. As shown in the hierarchical diagram in Fig. 6, level 1 represents absolute volumetric-based water allocation; level 2 represents time-based water allocation; while level 3 represents a percentage of the total flow. Levels 4 and 5 represent a combination of these various water allocation mechanisms. Transforming negotiation over water allocation means moving from level 1 to level 5 by reflecting the move from pure *division* to *sharing* of these resources.

By and large, time-based water allocation is more beneficial and desirable than volumetric-based allocation, although each mechanism is useful in certain cases. This is because time-based water allocation tends to distribute the risk of drought, caused by the unforeseen and unpredictable fluctuation of water availability in nature, evenly among co-riparians, instead of placing the risk of facing drought on one country. It also gives the countries flexibility to store water for times of shortage, providing high level of water security and incentives for conservatory policies. For the Palestinians and Israelis, this might suggest the need to consider utilizing regional storage assets, such as Lake Tiberias, for mutual storage needs to serve both countries. However, the parties must first reach agreement on basic water needs that in this particular case might call for a fixed and agreed upon minimum threshold of water volumes to safeguard the Palestinian domestic water rights. This is because volumetric-based allocation is most appropriate in responding to urgent and quantifiable needs, such as critical domestic water demands for the Palestinian communities. In the long term once these basic water needs are met, the parties might consider time-based water allocation, namely over agricultural or industrial water uses, to foster the stability and sustainability of the agreement, create a sense of ownership and responsibility towards the environment and boost cooperative relations among parties on other political matters.

It is crucial to note that in all of these three levels of water allocation, including volumetric-based, time-based, and percentage of total flow allocation, water resources are *divided* not *shared*. A far more sophisticated water allocation strategy can be realized when these approaches are combined, resulting in water *sharing* rather than *dividing*. While level 4 conjoins time-based and volumetric-based water allocation, level 5 is deemed the most desirable mechanism of water *sharing* for the Palestinians and Israelis as it conjoins time-based and percentages of total flow during wet and dry seasons. In short, *sharing* water between the Palestinians and Israelis means securing agreement on specific volumes of water for their most urgent and basic consumptive needs, while utilizing level 5 as the basis for long-term water negotiations to ensure security and sustainable cooperation.

4. **Needs-based water allocation**

The foundational principle of strategic water management is to acquire and transfer water where it is plentiful to where it is needed. In arid regions, such as the Middle East, water is perceived as life itself. The United Nations considers access to safe drinking water to be, like food, an indispensable basic human *right* and fundamental to life and health (United Nations Environment Programme 2002). Human water *right* in this context (or the right to water) generally includes abstracting, diverting, impounding or storing and using a specified quantity of water. It is distinguished from the legal connotation of water *rights* and ownership which is usually related to questions of sovereignty and riparian ownership of water resources which is often intertwined with issues related to land and international borders. Human water *right* means not only the ability to secure adequate water in terms of quantity and quality, but also the ease and reliability of access to that water to meet the basic level of human water *needs*. This *needs*-based water allocation should be the foundation for the negotiation between the Palestinians and Israelis. This recommendation to consider *needs* as the basis for water allocation in substitution for the argument of *rights* and entitlement emerges as a more effective and conducive mechanism to lead a win-win negotiation in the short term.

Given the ambiguity of the international law as well as the term water *needs* itself, the definition of the latter remains a crucial question to contend with. To avoid individual interpretation of the term water *needs*, a more objective approach is warranted where a neutral third party can define judicially cognizable *needs* based on demographic or economic grounds to estimate actual data that is measurable and agreeable. According to Wolf, 1999, water *needs* can be defined and quantified on the basis of catering to current and future population's water consumption, irrigable land, and requirements of specific projects. These water *needs* are universal human *needs* on a personal level regardless of how robust their society is in terms of economic status. This suggests utilizing a per capita needs-based allocation to quantify these water *needs*. This conclusion is also echoed by other scholars such as Gleick, 1994, who asserts that water should be allocated based on demographic grounds regardless of economic and other industrial needs. Reflecting interest-based rather than position-based negotiation, this recommendation seems to be in sync with the concept of prioritization of uses and users, where satisfying the urgent domestic and potable water *needs* is given higher priority than other industrial and agricultural uses, as we discussed earlier in this chapter. By focusing on judicially cognizable, identifiable *interests* and quantifiable *needs*, the parties will be able to mitigate the adverse impacts of entitlement that tend to dominate position-based negotiations and is likely to trigger a race to the bottom.

However, it is crucial to note that the concept of water *rights* follows the question of sovereignty over land and boarders. The Palestinians has a claim to their land that must be acknowledged and amicably resolved in these negotiations. Resolving the question of water *needs* in the short term cannot be viewed as a fulfillment or substitute for water and land *rights*. It is only a bridge to a short-term agreement that can offer water security to both nations, but without relinquishing the very essential and important question of land and sovereignty for the Palestinian people. In the spirit of incrementalism, a more comprehensive and final status negotiation must be able to address the inalienable rights of the Palestinians to self-determination including their *right* to land and water resources as part of their national identity and aspiration for independence.

5. **Satiety-based hydro-diplomacy**

As equitable agreements incorporate elements of both equitable *process* and equitable *outcomes*, one cannot neatly separate the two aspects of negotiations from one another. Although using elements of equable *process* in the treaty formation stage does not always guarantee equitable *outcomes*, relying on elements of equitable *process* enhances the chances of achieving equitable *outcomes* at the end. It also can be said that with the lack of equitable *process*, the prospect of reaching an equitable agreement is slim to nonexistent. I demonstrated in my previous book, *Water as a Catalyst for Peace,* that achieving equitable *outcomes* does not necessarily lead to high (positive) perceptions and acceptability of a certain treaty. This is because in cases where the conflict involves questions related to other issues such as political and cultural dimensions, high levels of hostility often result in a zero-sum game situation, where one riparian's gain is perceived as another's loss (Postel and Wolf 2001). Employing parameters of *process* equity tends to influence perception of a certain treaty as being equitable, which in turn drives the level of satisfaction and acceptability of its outcome. In the Palestinian-Israeli conflict, achieving *satiety* (which simply means satisfaction in the outcome of the agreement) for all parties is as crucial as the actual *outcome* of satisfying their very basic water *needs*. This is because in this particular context, it is pivotal to foster a win-win outcome, where both nations can satisfy their needs for water, while pari passu with this, perceive the other party as a partner in peace rather than an adversary in order to eliminate reasons for resentment and hostility. In short, *satiety* can be said to be primarily contingent upon the general *perception* of outcome, which includes the *perception* of the general public, *perception* of the impacted local community, and *perception* of the negotiators of the treaty. With this in mind, any future negotiations between the Palestinians and Israelis must address all of these types of perceptions to be successful in gaining acceptance and successful implementation. Of particular importance is the perceptions of the impacted local communities—those end users who bear the brunt of, or benefit from a certain water treaty. This segment of the public has their own unique incentives and reasons to believe and accept certain outcomes depending on how they impact their lives. On the other hand, the negotiators' perception of the treaty outcome is primarily shaped by their political motives and agenda, as well as non-water related gains, including strategic, political, and economic gains.

The question of representation resides at the very core of the issue of *satiety*. Given its poor track record in previous negotiation and the abominable reality it produced on the ground, the Palestinian Authority has lost any credibility and cannot be viewed as a representative of the Palestinian people. As we previously discussed in this book, a representative Palestinian leadership capable of rallying all Palestinians behind a united vision for peace is warranted to lead successful negotiations. By the same token, considering its extreme stance and full-throated opposition to a Palestinian state along with its expansive settlement policy, the current Israeli government also lost credibility as a viable partner in peace. Both nations are in need of a new leadership able to drive lessons and parallels from past mistakes in order to resurrect a viable vision for future peace. Moreover, tackling water in separation from the overall political question will not yield sustainable peace. As such, it is pivotal to address the question of water resources allocation in conjunction with other negotiation agenda items, namely economic, strategic, and diplomatic dimensions, as well as the question of land and border. The latter is particularly important in the context of achieving Palestinian independence and resolving the overall conflict and its underlying root causes that color almost every aspect of their relationship with Israel.

The gap between people's expectations of an agreement and its actual performance, which can diminish over time, also drives the overall perception of its outcome. Setting realistic expectations upfront through transparency about the scope and purpose of the negotiation process helps tremendously in mitigating this gap and can sway the public opinion towards a more acceptable perception and ultimately maintain acceptance and ensure successful implementation over time. This requires tight alignment between expectations and agreement performance over time, notwithstanding the fluctuation in both dimensions (performance of the agreement and the surrounding geopolitical landscape that influence both expectations and acceptance). In short, *satiety* emerges as a key ingredient in eliminating resentment and hostility and advancing negotiation to implementable and durable agreements. This is because the concept of *satiety* can be best described as recognizing one's needs, accepting one's share of water, and acknowledging the needs of others.

Proactive public involvement and the adoption of collaborative agreement-seeking processes are key components to achieving *satiety*. Proactive public involvement from the very onset of the negotiation process is key to building trust and eliminating reasons for doubts and suspicion. This can also create a platform for consensus building and agreement-seeking processes where parties and their constituents are aligned and supportive of the decisions and their outcomes, mitigating post settlement disputes. This can only be accomplished through constant collaboration and communication with the public about the treaty formation process and its outcomes throughout the entire negotiation lifecycle. Many of the previous negotiation attempts in the so called "peace process" and their ensuing ghastly work were surrounded by a great deal of secrecy. These clandestine agreements were spurned by both peoples due to this secrecy-induced mistrust that led to its demise and evident failure. A paradigm shift is needed here to move from the technocratic approach of negotiating on behalf of the people, known as "planning *for* people" to a more collaborative approach that focuses on "planning *with* people," by promoting a participatory and engaging negotiation process with high level of openness and transparency. This approach to negotiation necessitates moving away from the traditional approach of technocratic, expert-lead negotiation and decision making that depends heavily on the opinions of "water experts" to a more participatory approach grounded in community leadership, joint fact finding, and collaborative decision making. It is precisely the fact that eliminating the "duelling expert" syndrome can only transform the traditional negotiation to become more transparent, open, and collaborative, which in turn can provide a better institutional and social climate for implementation (Jarvis and Wolf 2010).

6. **Context sensitive equity and adaptive learning and management**

Serious dedication to fairness, which in far more aligned with the meaning of *equity*, rather than *equality*, is needed to move the needle on the current deadlock in the Palestinian-Israeli diplomatic relationship. In the context of the Palestinian-Israeli conflict, achieving equity and fairness is particularly important and calls for fair allocation of both land and water resources in keeping with the guidance of the international law. First, allocating land equitably simply means ending the occupation of Palestinian lands acquired in 1967, halting and dismantling illegal settlements, and ending the siege on Gaza to allow freedom of movement for the Palestinians. Second, water allocation between the two countries needs to guide the overall process to ensure that adequate water, in terms of quantity and quality, is allocated to the Palestinian communities, proportional to their actual needs, as previously discussed in this chapter. This means that an independent assessment of current water supply and population needs is warranted to ensure both sides agree on, and utilize, objective and accurate water data as a platform to build better agreements regarding more thorny issues. Attaining equitable water sharing between the two countries means distributing not only *benefits* but also *costs* and risks between them in a fashion that can foster acceptability and successful implementation. Sharing benefits means allocating comparable amounts of water per capita for both nations while sharing cost means distributing risks of droughts evenly between them.

Since water availability and population needs are both in flux, the concept of equity need not be rigid. Resilience in water allocation over time must be incorporated to respond to situational changes. Incorporating an ongoing situational assessment of water demand and supply and pivoting water allocation accordingly is needed to keep up with the ever-changing nature of water. Negotiators need to be cognizant of this crucial factor as they consider different water allocation alternatives to ensure flexibility and adaptability is possible, while at the same time, accountability is achieved and maintained. This is of course tricky and requires a system of ongoing cooperation and dialog, which not only can ensure resilience but can also maintain sustainable channels of communications, especially after a deal is reached. Due to the fact that international relationship in the context of the Palestinian-Israeli conflict is based on asymmetry of interests, the parties must focus their attention and effort on enabling a mutual agreement whereby each side can realize their respective "win" without having the other side "lose." This requires the Palestinians and Israelis to tackle multiple issues, such as their differential water needs, political relationships, and non-water related issues, in order to create a common platform to trade across the various value systems and enable them to augment their respective net gain.

Water issues, on the other hand, are complex in nature cutting across a multitude of social, natural and political domains and morph into new realities and challenges. This complexity is a byproduct of the interaction between numerous dimensions related to processes, people, and institutions at various levels and scales (Islam and Susskind 2013). The boundary-crossing nature of water calls for a critical understanding and careful examination of the nuances and intricacies of each individual case, as each case is unique and the Palestinian-Israeli conflict is no exception in this sense. Over time, the Palestinian-Israeli water conflict changed along four aspects, including conflict formation (or emergence of new disputes), conflict exacerbation, conflict mitigation, and conflict resolution or transformation (Mitchell 2006). Water conflicts and their resolutions change over time and so should agreements. As such, any future agreement should encapsulate elements of flexibility to adapt to new situations and challenges in order to provide an adaptable "resolutionary" platform to transform this protractible conflict to a more stable and plausible state. Critical to this is the ability to strike the right balance between maintaining flexibility and ensuring compliance in the terms of the agreement. When compliance suffers, flexibility must cease to allow both sides to deliver on their commitments as per the agreement. With this in mind, flexibility must not be confused with holding each other accountable for successful implementation of the terms of the agreement, a situation that prevailed in previous negotiations between the two sides as discussed in the previous chapter.

7. **Linkage to environmental and non-environmental issues**

Environmental issues, such as water conservation and rights of the environment, and non-environmental issues, such as political, strategic, and economic factors, are usually intertwined in water negotiations. They must be negotiated in tandem to achieve win-win solutions through prioritization and trade-offs. Advocating for the right of the environment calls for a fundamental change in the way the Palestinian-Israeli negotiations are managed in order to view the environment as a player in need of its fair share of water, rather than a commodity to be concurred, exploited and divided. This is where the so called "peace process" fell short, as Israel continued to extract significant volumes of the Jordan River water to the point where the integrity and health of the Dead Sea ecosystem is now compromised. To achieve fairness in future deals, both sides must be cognizant of the importance of balancing human needs with environmental concerns in order to mitigate environmental degradation to the shared water resources. On the other hand, non-environmental issues, such as political, strategic, and economic gains must also be considered as key elements to achieving a successful negotiation process. Each party value these aspects differently and as such achieving mutual gains and carving out acceptable ZOPA (Zone of Possible Agreement) can be possible when these aspects are considered in their entirety. For example, Israel often value cooperation and regional and political stability as a beneficial outcome from reaching an agreement, whereas the Palestinians view regaining control over their land and sovereignty as an acceptable outcome. Negotiating these aspects together enhances the parties' ability to find workable solutions and move their negotiation from a zero-sum game to positive-sum game where both sides can realize their respective important goals in a negotiated settlement.

8. **Goodwill/faith-based confidence building measures (CBMs)**

Goodwill on both sides prior to entering into negotiation is a crucial element to incentivize the negotiating parties to cooperate and ameliorate contentious issues that otherwise would not be easily resolved in a conventional manner. The previous negotiation attempts that the Palestinians and Israelis led failed in this regard as they mainly focused on entitlement issues and with the presence of mistrust. This unfortunate situation resulted in a stalemate, where both parties are put on a collision course with each other.

Bridging the cultural and communication gap prior to entering negotiation means building trust in each other as partners in peace and confidence in the process by separating the problem from the people, focusing on broader and identifiable common goals, and eliminating the underlying reasons of tension and conflict, namely settlement expansion. To that end, building trust and demonstrating good faith suggests that Israel must end its occupation of the Palestinian land, halt its settlements, and end the siege on Gaza, all of which will undoubtedly have tremendous positive impacts on the negotiation process and the prospects of achieving real peace and stability in this region. In addition to being legal obligations, pulling back to the 1967 borders, ending and dismantling settlements, and giving the Palestinians their right to self-determination are all good steps not only for the Palestinians but also for Israel to signal a serious commitment to peace and the aptitude towards resolving their conflict once and for all.

9. **Incorporating modern and indigenous approaches to conflict resolution**

Due to the divergence in both history and cultural backgrounds, a one-sided approach to negotiation will be challenged on the grounds of being perceived as a western panacea attempting to reinforce control over the region and its resources. As was the case in previous negotiation attempts that failed to acknowledge local cultural practices, the results are predictable and will only lead to widening the cultural and communication divide. A more appropriately attuned approach to negotiation between the Palestinians and Israelis is one that incorporates common elements from modern and indigenous practices that fortunately are not antithetical in practical sense. This way, the aforementioned divide can be bridged and the negotiation process and its outcome will have a better chance of acceptance and support from both sides. The unspoken, yet well-known, set of principles that governs the societal interaction and community discourse in relation to resolving water disputes provides excellent tools to utilize during the negotiation process to enhance the acceptability of agreements and the implementation of their outcomes. Although helpful in fostering acceptability, combining modern and indigenous practices of conflict resolution was not why previous negotiations failed in reaching an acceptable deal, considering the nature of the dispute whose roots are related to geopolitical questions, such as sovereignty and independence that were not adequately resolved.

2.3 *Post-negotiation*

Reaching a deal is only the beginning of another important stage of the negotiation and treaty formation lifecycle. The real test of success of any agreement lies not in its novelty or cleverness, but rather in its success in sustaining implementation. While negotiation helps in defining success, the prost-negotiation stage helps in measuring it by ensuring successful adherence and compliance. Monitoring of the implementation of the agreement plays a critical role in gaging whether the parties are able to live up to their commitment and if the agreement is able to yield the desired outcome it was designed to accomplish, in relation to the political, economic and environmental aspects. It also unearths execution and implementation hurdles that the parties need to figure out together to overcome. In service of these vital goals, assessing, facilitating, and overseeing implementation and enticing all parties to maintain continuous dialog and cooperative relationship with each other beyond the initial agreement is crucial to lead successful implantation. In addition, in this stage, there might be a need to conduct post-settlement negotiation to iron out implementation details or other issues that can potentially arise as a result of new circumstances and challenges. Flexibility is needed, so is the ability to keep commitments and ensure compliance. As such, in the post-negotiation stage, the parties must continue to work together and create a cadence of cooperation and continued dialog. Enhancing basin-wide cooperation, coalition building, and partnership are key components in the establishment of a fluent institutional structure necessary to monitor implementation while providing resilience and adaptability to new changes and the ambiguity, complexity, and long-term consequences of water allocation decisions.

This stage also provides a platform to demonstrate the value of ratified agreements. Beyond political and partisan rhetoric, the parties have greater responsibility to provide transparency into the agreement performance and its actual outcome in a measurable, quantifiable, and demonstrable fashion. This is because agreements are designed to last for a specific period of time after which point they either expire or require new ones to be put in place to address the issues at hand. Understanding when new agreements are deemed necessary is a pivotal goal of the post-negotiation stage. This can be achieved by continuing to assess the agreement performance along two crucial dimensions, including the perceived benefit of the agreement outcomes, which tends to diminish over time, and the extent to which the parties adhere to its provisions. Perceptions of benefit determine whether the resolution is acceptable or unacceptable, and are shaped by the gap between what people expect and what they actually receive as a result of post-negotiation agreement performance. Implementation can be impacted and potentially compromised when expectations exceed the actual agreement outcome as manifested in its performance. This is the point where the parties need to look at revising the agreement or introducing a new one to narrow the gap between people's expectations and its actual performance. This warrants the establishment of continuous channels of communication and dialog; the creation of institutional designs (such as joint water committees) conducive to govern the implementation phase, ensure compliance, and measure success; and most importantly the willingness to pivot course when the agreement outlives its perceived benefits. It all comes down to the parties' ability to continue the dialog, which requires the wisdom to *listen* and credibility and courage to *lead*, both of which are lacking in the current political atmosphere.

Third: Neutral Third-Party Mediation Process

By definition, mediation is where an impartial third party amicably facilitates communication among disputants to promote reconciliation, settlement, or agreement. The involvement of a neutral third-party mediator is instrumental in many ways and cannot be overemphasized. In fact, the evident failure of previous negotiations between the Palestinians and Israelis can be largely attributed to the lack of a neutral mediator, as discussed earlier in the book. The mediator can play a critical role not only in incentivizing the parties to come to the negotiation table, but also in conveying to the two sides that the BATNA[2] is equally deplorable and that walking away

[2] Best alternative to a negotiated agreement.

from the negotiation table is not an option. A major step in cementing relationships based on trust should take place prior to negotiation with the help and guidance of a neutral third-party. This is particularly important in uniting both nations to endorse a common rubric for peace and reconciliatory processes. The mediator also helps in identifying practical negotiation agenda, common interests and gaps, generating and evaluating options, removing obstacles, and prompting implementation. In this regard, the involvement of a neutral mediator is instrumental during the negotiation process in keeping the parties "honest" by neutralizing the impact of power structure imbalance, communication problems, misrepresentation, and misinformation (SPIDR 1997). Once a deal is reached, it also helps in soliciting and securing economic assistance to incentivize the parties to reach a negotiated settlement and fund its implementation to foster better economic conditions, especially for the Palestinian communities suffering the most as a result of this conflict.

Neutrality in mediation, however, is paramount especially in the context of the Palestinian-Israeli conflict, which witnessed first-hand the detrimental impacts of the U.S. involvement in the process in the capacity of mediating a peace process that was doomed to fail from the very beginning due to the impartiality of the mediator. This was evident in the strategic blunder that both parties found themselves in as a result of taking part in this undertaking. As discussed in previous chapters, under the aegis of the U.S., the process ended up serving as an assiduous diplomatic cover to allow Israel to expand its control while deny the Palestinians their basic human rights, including the right to water and land (Khalidi 2013). Consequently, the U.S. lost its credibility as a mediator in this conflict, as the promise of achieving peace remained unfulfilled. This vacuum that was created as a result of the aforementioned lack of objectivity and impartiality exhibited by the U.S. involvement calls for a new even-handed, non-allied and neutral third-party mediator in any future negotiations in order to achieve an equidistant approach to negotiation capable of bridging the communication gap in order to reach a plausible, acceptable, and just peace settlement. This lack of a neutral third-party mediator created a vacuum that constitutes a serious threat to future negotiations between the Palestinians and Israelis. It remains to be seen whether the parties will be able to identify a mediator that would advocate for a united vision for peace and reinforce the path to a just and sustainable peace settlement in order to transform the conflict and advance negotiations to a more amicable state.

Afterthought and Closing Remarks

Paradoxical as it may seem, any discussion about Middles East peace or stability inevitably turns into a discussion about Palestine (Said 2003). As the land of Palestine went through a plethora of transformations, both in terms of its geographic configurations as well as its demographic characteristics, the Palestinian identity remained and even evolved stronger with time. Uprootedness, dispossession and the doubly and triply miserable faith of the Palestinians shaped their experience, sense of identity and political solidarity that remained so unique and powerful to this very day. Victimhood and the feeling of loss of one's own land that many Palestinians share with their refugee compatriots and all of those who had to endure living under occupation for decades brings them together under one umbrella—Palestine. The unwavering universal Palestinian desire for self-determination acquired unbudgeable place in the world's agenda for freedom, human rights, and sustainability. This is clearly reflected by the ongoing emphasis of many international organizations, endeavors, and activities that showcase the Palestinian cause as a protagonist epitomizing all of these values. There is a strong moral and political solidarity that vibrate throughout the world and unify the Palestinians with a remarkable sense of belonging. This potent sense of belonging was always encumbered by memories and stories about Palestine narrated by their parents and grandparents who survived the *Nakba*. The diverse and pervasive affirmation of the Palestinian identity brought to existence a remarkable and healthy awareness of the ironies of unequal powers, despite the world's indifference (Said 2003). The more Israel unilaterally intensifies its considerable effort to erase and burry the Palestinian past, a painstaking collective effort, albeit unorganized, to preserve the Palestinian memories persists generations after generations. This is what kept Palestine alive.

Historically Palestine has been home for many civilizations that serially inhabited the land. To say that one group of people can only lay claim to this land is untenable. Edward Said, demonstrating this rich historical and cultural character of Palestine and advocating for coexistence explains that "Palestine is a palimpsest of an enormous number of peoples and civilizations each of them serially leaving traces and effects that have lasted for several centuries… To extrapolate from that

A. Abukhater, *Palestine - Peace by Piece*,
https://doi.org/10.1007/978-3-030-03985-1

and say that only one people dominated it is therefore entitled to lay exclusive claim to Palestine today is simply a tragic misinterpretation and betrayal of something far richer and interesting. One could interpret that history therefore inclusively giving rise to complexity and see universalism or far more narrowly as the exclusive possession of only one people in culture who seem to prefer unending war against others to mutual recognition and coexistence." (Said 2003). This vision for coexistence must be endorsed by both people to arrive to a just and lasting peace that enable both nations to live side by side with full independence and security.

As we have seen, many solutions have perennially been proposed to settle the dispute, none was mature or comprehensive enough to make any substantive improvement in the lives of Palestinians inside the territories or in diaspora whose rights have been scandalously violated for far too many decades to count. These solutions, by design, were meant to minimize and cast aside the issue of Palestinian refugees in an attempt to silence or marginalize their voice. These solutions, *ipso facto*, did not succeed. On the contrary, they exacerbated an already volatile situation, and led to too many deaths and miseries to count. The pressure that has been building is a byproduct of the so called "peace process" that resulted in a situation where the Palestinians feel no meaningful resolution has been reached. Throughout the years following the Oslo agreements, things went from bad to worse. More land appropriation, more settlement, siege and multiple wars resulted in an isolated Gaza and fragmented West Bank.

As a beleaguered outnumbered minority of Palestinians and Israelis sought peace and reconciliation to settle the conflict, the issue remains however that the majority of Israelis and Palestinians still needs to truly and honestly endorse this option as a viable alternative to frustration and violence, beyond lip service. Unless both nations are mobilized to get behind this peaceful option, the situation will continue and the cycle of violence will turn into an eternal conflict. Despite the current situation, the human suffering, and the stalemate in the peace process that misled both nations, I remain hopeful that a more peaceful ending could be possible. For any future solution to have a chance to succeed, it must recognize the inalienable rights of the Palestinian refugees to return to their homeland from which they were forced out in 1948. Only then can we see the light at the end of the tunnel. To give peace a chance, Israel must abide by its legal obligation to dismantle its illegal settlements and end its occupation and siege on Gaza with a granitic well to genuinely achieve peace and a prevailing spirit of contrition over the years of occupation and dispossession. A genuine attempt with the art of the possible to resurrect a unified vision for peace by both nations that realizes the rights of each other and the obligation for one another is badly and urgently needed to rectify and *mend* what's been *broken*—one piece at a time.

Appendix A: List of Palestinian Villages Depopulated and Destroyed in 1948 *Nakba* (Catastrophe)

The District of Acre—26 villages
Amqa, Arab al-Samniyya, al-Bassa, al-Birwa, al-Damun, Dayr al-Qasi, al-Gabisiyya, Iqrit, Khirbat Iribbin, Khirbat Jiddin, al-Kabri, Kufr Inan, Kuwaykat, al-Manshiyya, al-Mansura, Miar, al-Nabi Rubin, al-Nahr, al-Ruways, Suhmata, al-Sumayriyya, Suruh, al-Tall, Tarbikha, Umm al-Faraj, al-Zib.

The District of Bisan—29 villages
Arab al-'Arida, Arab al-Bawati, Arab al-Safa, al-Ashrafiyya, al-Birra, Danna, Farwana, al-Fatur, al-Ghazzawiyya, al-Hamidiyya, al-Hamra, Jabbul, Kafra, Kawkab al-Hawa, Arab al-Khunayzir, Masil al-Jizl, al-Murassas, Qumya, al-Sakhina, al-Samiriyya, Sirin, Tall al-Shawk, Khirbat al-Taqa, al-Tira, Umm 'Ajra, Umm Sabuna, Yubla, Zab'a, Khirbat al-Zawiya.

The District of Beersheba—3 villages
Al-'Imara, al-Jammama, al-Khalasa

The District of Gaza—45 villages
Arab Suqrir, Barbara, Barqa, al-Batani al-Gharbi, al-Batani al-Sharqi, Bayt 'Affa, Bayt Daras, Bayt Jirja, Bayt Tima, Bi'lin, Burayr, Dayr Sunayd, Dimra, al-Faluja, Hamama, Hatta, Hiribya, Huj, Hulayqat, 'Ibdis, 'Iraq al-Manshiyya, Iraq Suwaydan, Isdud, al-Jaladiyya, al-Jiyya, Julis, al-Jura, Jusayr, Karatiyya, Kawfakha, Kawkaba, al-Khisas, al-Masmiyya al-Kabira, al-Masmiyya al-Saghira, al-Muharraqa, Najd, Ni'ilya, Qastina, al-Sawafir al-Gharbiyya, al-Sawafir al-Shamaliyya, al-Sawafir al-Sharqiyya, Simsim, Summayl, Tall al-Turmus, Yasur.

The District of Haifa—51 villages
Abu Shusha, Abu Zurayq, Arab al-Fuqara', Arab al-Nufay'at, Arab Dhahrat al-Dhumayri, Atlit, Ayn Ghazal, Ayn Hawd, Balad al-Shaykh, Barrat Qisarya, Burayka, Khirbat al-Burj, Khirbat al-Butaymat, Daliyat al-Rawha', Khirbat al-Damun, Khirbat al-Ghubayya al-Fawqa, Khirbat al-Ghubayya al-Tahta, Hawsha, Ijzim, Jaba', al-Jalama, Kabara, al-Kafrayn, Kafr Lam, Khirbat al-Kasayir, Khubbayza, Khirbat Lid, Khirbat al-Manara, al-Mansi, Khirbat al-Mansura,

A. Abukhater, *Palestine - Peace by Piece*,
https://doi.org/10.1007/978-3-030-03985-1

al-Mazar, al-Naghnaghiyya, Qannir, Qira, Qisarya, Qumbaza, al-Rihaniyya, Sabbarin, al-Sarafand, Khirbat al-Sarkas, Sa'sa', Khirbat al-Sawamir, Khirbat al-Shuna, Khirbat al-Sindiyana, al-Tantura, al-Tira, Umm al-Shawf, Umm al-Zinat, Wa'arat al-Sarris, Wadi Ara, Yajur.

The District of Hebron—16 villages
'Ajjur, Barqusiya, Bayt Jibrin, Bayt Nattif, al-Dawayima, Dayr al-Dubban, Dayr Nakhkhas, Kidna, Mughallis, al-Qubayba, Ra'na, Tall al-Safi, Khirbat Umm Burj, Zakariyya, Zayta, Zikrin.

The District of Jaffa—23 villages
al-'Abbasiyya, Abu Kishk, Bayt Dajan, Biyar 'Adas, Fajja, al-Haram, Ijlil al-Qibliyya, Ijlil al-Shamaliyya, al-Jammasin al-Gharbi, al-Jammasin al-Sharqi, Jarisha, Kafr 'Ana, al-Khayriyya, al-Mas'udiyya, al-Mirr, al-Muwaylih, Rantiya, al-Safiriyya, Salama, Saqiya, al-Sawalima, al-Sheikh Muwannis, Yazur.

The District of Jerusalem—38 villages
'Allar, 'Aqqur, 'Artuf, 'Ayn Karim, Bayt 'Itab, Bayt Mahsir, Bayt Naqquba, Bayt Thul, Bayt Umm al-Mays, al-Burayi, Dayr Aban, Dayr 'Amr, Dayr al-Hawa, Dayr Rafat, Dayr al-Shaykh, Dayr Yassin, Ishwa', Islin, Khirbat Ism Allah, Khirbat Jarash, al-Jura, Kasla, Khirbat al-Lawz, Lifta, al-Maliha, Nitaf, al-Qabu, Qalunya, al-Qastal, Ras Abu 'Ammar, Sar'a, Saris, Sataf, Suba, Sufla, Khirbat al-'Umur, al-Walaja.

The District of Jenin—6 villages
Ayn al-Mansi, Khirbat al-Jawfa, al-Lajjun, al-Mazar, Nuris, Zir'in.

The District of Nazareth—4 villages
Indur, Ma'lul, al-Mujaydil, Saffuriyya.

The District of Ramla—58 villages
Abu al-Fadl, Abu Shusha, 'Ajanjul, 'Aqir, Barfiliya, al-Barriyya, Bashshit, Khirbat Bayt Far, Bayt Jiz, Bayt Nabala, Bayt Shanna, Bayt Susin, Bir Ma'in, Bir Salim, al-Burj, Khirbat al-Buwayra, Daniyal, Dayr Abu Salama, Dayr Ayyub, Dayr Muhaysin, Dayr Tarif, Khirbat al-Duhayriyya, al-Haditha, Idnibba, 'Innaba, Jilya, Jimzu, Kharruba, al-Khayma, Khulda, al-Kunayyisa, al-Latrun, al-Maghar, Majdal Yaba, al-Mansura, al-Mukhayzin, al-Muzayri'a, al-Na'ani, al-Nabi Rubin, Qatra, Qazaza, al-Qubab, Qubayba, Qula, Sajad, Salbit, Sarafand al-'Amar, Sarafand al-Kharab, Saydun, Shahma, Shilta, al-Tina, al-Tira, Umm Kalkha, Wadi Hunayn, Yibna, Khirbat Zakariyya, Zarnuqa.

The District of Safad—77 villages
Abil al-Qamh, al-Abisiyya, Akbara, Alma, Ammuqa, 'Arab al-Shamalina, Arab al-Zubayd, 'Ayn al-Zaytun, Baysamun, Biriyya, al-Butayha, al-Buwayziyya, Dallata, al-Dawwara, Dayshum, al-Dhahiriyya al-Tahta, al-Dirbashiyya, al-Dirdara, Fara, al-Farradiyya, Fir'im, Ghabbatiyya, Ghuraba, al-Hamra', Harrawi, Hunin, al-Husayniyya, Jahula, al-Ja'una, Jubb Yusuf, Kafr Bir'im, al-Khalisa, Khan al-Duwayr, Khirbat Karraza, al-Khisas, Khiyam al-Walid, Kirad al-Baqqara, Kirad

al-Ghannama, Lazzaza, Madahil, al-Malikiyya, Mallaha, al-Manshiyya, al-Mansura, Mansurat al-Khayt, Marus, Mirun, al-Muftakhira, Mughr al-Khayt, Khirbat al-Muntar, al-Nabi Yusha', al-Na'ima, Qabba'a, Qadas, Qaddita, Qaytiyya, al-Qudayriyya, al-Ras al-Ahmar, Sabalan, Safsaf, Saliha, al-Salihiyya, al-Sammu'i, al-Sanbariyya, Sa'sa, al-Shawka al-Tahta, al-Shuna, Taytaba, Tulayl, al-'Ulmaniyya, al-'Urayfiyya, al-Wayziyya, Yarda, al-Zanghariyya, al-Zawiya, al-Zuq al-Fawqani, al-Zuq al-Tahtani.

The District of Tiberias—25 villages

'Awlam, al-Dalhamiyya, Ghuwayr Abu Shusha, Hadatha, al-Hamma, Hittin, Kafr Sabt, Lubiya, Ma'dhar, al-Majdal, al-Manara, al-Manshiyya, al-Mansura, Khirbat Nasser al-Din, Nimrin, al-Nuqayb, Samakh, al-Samakiyya, al-Samra, al-Shajara, al-Tabigha, al-'Ubaydiyya, Wadi al-Hamam, Khirbat al-Wa'ra al-Sawda',Yaquq.

The District of Tulkarem—17 villages

Khirbat Bayt Lid, Bayyarat Hannun, Fardisya, Ghabat Kafr Sur, al-Jalama, Kafr Saba, al-Majdal, Khirbat al-Manshiyya, Miska, Qaqun, Raml Zayta, Tabsur, Umm Khalid, Wadi al-Hawarith, Wadi Qabbani, Khirbat al-Zababida, Khirbat Zalafa.

The list and figures are taken from: "All That Remains: The Palestinian Villages Occupied and Depopulated by Israel in 1948", Ed. Walid Khalidi. Berkeley: University of California Press, 1991.

Appendix B: Full Transcript of the U.S. Secretary of State John Kerry's Speech on Dec. 28, 2016

Presenting Six Points for Future Peace Deal and Outlining the U.S. Position Towards Resolving the Palestinian-Israeli Conflict

"Thank you all. It's good to be here even in the middle of a holiday week. I wish you all a happy and productive new year. Today, I want to share candid thoughts about an issue that for decades has animated the foreign policy dialogue here and around the world—the Israeli-Palestinian conflict. Throughout his administration, President Obama has been deeply committed to Israel and its security, and that commitment has guided his pursuit of peace in the Middle East. This is an issue which I've worked on intensely during my time as Secretary of State for one simple reason: because the two state solution is the only way to achieve a just and lasting peace between Israelis and Palestinians. It is the only way to ensure Israel's future as a Jewish and democratic state, living in peace and security with its neighbors. It is the only way to ensure a future of freedom and dignity for the Palestinian people. And it is an important way of advancing U.S. interests in the region. I would like to explain why that future is now in jeopardy, and provide some context for why we could not, in good conscience, stand in the way of a resolution at the United Nations that makes clear that both sides must act now to preserve the possibility of peace. I am also here to share my conviction that there is still a way forward if the responsible parties are willing to act. And I want to share practical suggestions for how to preserve and advance the prospects for the just and lasting peace that both sides deserve. It is vital that we have an honest, clear-eyed conversation about the uncomfortable truths and difficult choices, because the alternative that is fast becoming the reality on the ground is in nobody's interest—not the Israelis, not the Palestinians, not the region—and not the United States. I want to stress that point:

A. Abukhater, *Palestine - Peace by Piece*,
https://doi.org/10.1007/978-3-030-03985-1

My job, above all, is to defend the United States of America—to stand up for and defend our values and our interests in the world. If we were to stand idly by and know that in doing so we are allowing a dangerous dynamic to take hold which promises greater conflict and instability to a region in which we have vital interests, we would be derelict in our own responsibilities. Regrettably, some seem to believe that the US friendship means the US must accept any policy, regardless of our own interests, our own positions, our own words, our own principles—even after urging again and again that the policy must change. Friends need to tell each other the hard truths, and friendships require mutual respect. Israel's permanent representative to the United Nations, who does not support a two-state solution, said after the vote last week: It was to be expected that Israel's greatest ally would act in accordance with the values that we share and veto this resolution. I am compelled to respond that the United States, did in fact vote "in accordance with our values," just as previous U.S. administrations have done at the Security Council. They fail to recognize that this friend, the United States, that has done more to support Israel than any other country, this friend that has blocked countless efforts to delegitimize Israel, cannot be true to our own values—or even the stated democratic values of Israel—and we cannot properly defend and protect Israel—if we allow a viable two state solution to be destroyed before our eyes. And that's the bottom line: the vote in the UN was about preserving the two state solution. That's what we were standing up for: Israel's future as a Jewish and democratic state, living side by side in peace and security with its neighbors. That's what we are trying to preserve, for our sake and for theirs. In fact, this administration has been Israel's greatest friend and supporter, with an absolutely unwavering commitment to advancing Israel's security and protecting its legitimacy. On this point, I want to be very clear. No American Administration has done more for Israel's security than Barack Obamas. The Israeli Prime Minister himself has noted our unprecedented military and intelligence cooperation. Our military exercises are more advanced than ever. Our assistance for Iron Dome has saved countless Israeli lives. We have consistently supported Israel's right to defend itself, by itself, including during actions in Gaza that sparked great controversy. Time and again we have demonstrated that we have Israel's back. We have strongly opposed boycotts, divestment campaigns and sanctions targeting Israel in international fora, whenever and wherever its legitimacy was attacked, and we have fought for its inclusion across the UN system. In the midst of our own financial crisis and budget deficits, we repeatedly increased funding to support Israel. In fact, more than half of our entire global Foreign Military Financing goes to Israel. And this fall, we concluded an historic 38 billion dollar Memorandum of Understanding that exceeds any military assistance package the U.S. has provided to any country, at any time, and that will invest in cutting edge missile defense, and sustain Israel's Qualitative Military Edge for years to come. This commitment to Israel's security is very personal for me. On my first trip to Israel as a young senator in 1986, I was captivated by a special country I immediately admired and soon grew to love. Over the years, like so many others who are drawn to this extraordinary place, I have climbed Masada, swum in the Dead Sea, driven from one Biblical city to another. I have also seen the dark side of

Hezbollah's rocket storage facilities just across the border in Lebanon, walked through exhibits on the hell of the Holocaust at Yad Vashem, stood on the Golan Heights, and piloted an Israeli jet over the tiny airspace of Israel, which would make anyone understand the importance of security to Israelis. Out of those experiences came a steadfast commitment to Israel's security that has never wavered for a single minute in my 28 years in the Senate or my 4 years as Secretary.

I have also often visited West Bank communities, where I met Palestinians struggling for basic freedom and dignity amidst the occupation, passed by the military checkpoints that can make even the most routine daily trips to work or school an ordeal, and heard from business leaders who could not get the permits needed to get their products to the market and families who have struggled to secure permission to travel for needed medical care. And I have witnessed first-hand the ravages of a conflict that has gone on for far too long. I've seen Israeli children in Sderot whose playgrounds had been hit by Katyusha rockets, and visited shelters next to schools in Kiryat Shmona that kids had 15 s to get to after a warning siren went off. I've also seen the devastation of war in the Gaza Strip, where Palestinian girls in Izbet Abed Rabo played in the rubble of a bombed-out building. No children—Israeli or Palestinian—should have to live like that. So, despite the obvious difficulties, I knew when I became Secretary of State I would do everything in my power to help end the conflict. And I was grateful to be working for President Obama, who was prepared to take risks for peace and was deeply committed to that effort. Like previous U.S. administrations, we have committed our influence and resources to trying to resolve the Arab-Israeli conflict because it would serve American interests to stabilize a volatile region and fulfil America's commitment to the survival, security and well-being of Israel at peace with its Arab neighbors. Despite our best efforts over the years, the two state solution is now in serious jeopardy. The truth is that trends on the ground—violence, terrorism, incitement, settlement expansion and the seemingly endless occupation—are destroying hopes for peace on both sides and increasingly cementing an irreversible one-state reality that most people do not actually want. Today, there are a similar number of Jews and Palestinians living between the Jordan River and the Mediterranean Sea. They have a choice. They can choose to live together in one state, or they can separate into two states. But here is a fundamental reality: if the choice is one state, Israel can either be Jewish or democratic—it cannot be both—and it won't ever really be at peace. Moreover, the Palestinians will never fully realize their vast potential in a homeland of their own with a one state solution. Most on both sides understand this basic choice, and that's why it's important that polls of Israelis and Palestinians show there is still strong support for the two state solution—in theory. They just don't believe that it can happen. After decades of conflict, many no longer see the other side as people, only as threats and enemies. Both sides continue to push a narrative that plays to people's fears and reinforces the worst stereotypes—rather than working to change perceptions and build up belief in the possibility of peace. And the truth is, the extraordinary polarization in this conflict extends beyond Israelis and Palestinians. Allies of both sides are content to reinforce this with us or against us mentality where too often anyone who questions Palestinian actions is an apologist for the occupation and anyone who disagrees with Israeli policy is cast as anti-Israel or even anti-Semitic. That's one of the most striking realties about the current situation: This critical decision about the future—one state or two states—is effectively being made on the ground every day, despite the expressed opinion of the majority of the people. The status quo is leading towards one state and perpetual occupation, but most of the public either ignores it or has given up hope that anything can be done to change it. With this passive resignation, the problem only gets worse, the risks get greater and the choices are narrowed. This sense of hopelessness among Israelis is exacerbated by the continuing violence, terrorist attacks against civilians and incitement—which are destroying belief in the possibility of peace. Let me say it again: There is absolutely no justification for terrorism, and there never will be. The most recent wave of Palestinian violence has included hundreds of terrorist attacks in the past year, including stabbings, shootings, vehicular attacks and bombings, many by individuals who have been radicalized by social media. Yet the murderers of innocents are still glorified on Fatah web sites, including showing attackers next to Palestinian leaders following attacks. And despite statements by President Abbas and his party's leaders making clear their opposition to violence, too often they send a different message by failing to condemn specific terrorist attacks and naming public squares, streets and schools after terrorists. President Obama and I have made clear to the Palestinian leadership countless times, publicly and privately, that all incitement to violence must stop. We have consistently condemned violence and terrorism—and even condemned the Palestinian leadership for not condemning it. Far too often, the Palestinians have pursued efforts to delegitimize Israel in international fora. We have strongly opposed these initiatives, including the recent wholly unbalanced and inflammatory UNESCO resolution regarding Jerusalem. And we have made clear our strong opposition to Palestinian efforts against Israel at the ICC, which only set back the prospects for peace. And we all understand that the Palestinian Authority also has more to do to strengthen institutions and improve governance. Most troubling of all, Hamas continues to pursue an extremist agenda: they refuse to accept Israel's very right to exist. They have a one state vision of their own: all of the land is Palestine. Hamas and other radical factions are responsible for most explicit forms of incitement to violence, and many of the images they use are truly appalling. And they are willing to kill innocents in Israel and put the people of Gaza at risk in order to advance that agenda. Compounding this, the humanitarian situation in Gaza, exacerbated by the closures of the crossings, is dire. Gaza is home to one of the world's densest concentrations of people enduring extreme hardships with few opportunities. 1.3 million people out of Gaza's population of 1.8 million are in need of daily assistance, including food and shelter, most have electricity less than half the time, and only 5% of the water is safe to drink.

And yet despite the urgency of these needs, Hamas and other militant groups continue to re-arm and divert reconstruction materials to build tunnels, threatening more attacks on Israeli civilians that no government can tolerate. At the same time, we must be clear about what is happening in the West Bank. The Israeli Prime Minister publicly supports a two state solution, but his current coalition is the most right wing in Israeli history, with an agenda driven by its most extreme elements. The result is that policies of this government—which the Prime Minister himself just described as more committed to settlements than any in Israel's history—are leading in the opposite direction, towards one state. In fact, Israel has increasingly consolidated control over much of the West Bank for its own purposes—effectively reversing the transition to greater Palestinian civil authority called for by the Oslo accords. I don't think most people in Israel—and certainly in the world—have any idea how broad and systematic this process has become. The facts speak for themselves. The number of settlers in the roughly 130 Israeli settlements east of the 1967 lines has steadily grown. The settler population in the West Bank alone—not including East Jerusalem—has increased by nearly 270,000 since Oslo, including 100,000 just since 2009 when President Obamas term began. And there is no point pretending they're just in large settlement blocs: nearly 90,000 settlers are living east of the separation barrier that was created by Israel itself, in the middle of what by any reasonable definition would be the future Palestinian state. And the population of these distant settlements has grown by 20,000 just since 2009. In fact, just recently the government approved a significant new settlement well east of the barrier—closer to Jordan than Israel. What does that say to Palestinians in particular —but also to the U.S and the world—about Israel's intentions? Let me emphasize: this is not to say that the settlements are the whole or even primary cause of the conflict—of course they are not. Nor can you say that if they were removed you would have peace without a broader agreement—you would not. And we understand that in a final status agreement, certain settlements would become part of Israel to account for the changes that have taken place over the last 49 years, including the new demographic realities on the ground. But if more and more settlers are moving into the middle of the Palestinian areas, its going to be that much harder to separate, that much harder to imagine transferring sovereignty—and that is exactly the outcome that some are accelerating. Let's be clear: settlement expansion has nothing to do with Israel's security; many settlements actually increase the security burden on the IDF. And leaders of the settler movement are motivated by ideological imperatives that entirely ignore legitimate Palestinian aspirations.

Among the most troubling illustrations of this point has been the proliferation of settler outposts that are illegal under Israel's own laws. They are often located on private Palestinian land and strategically placed to make two states impossible. There are over 100 of these outposts, and since 2011, nearly one third have been—or are being—legalized, despite pledges by past Israeli governments to dismantle many of them. Now, leaders of the settler movement have advanced unprecedented new legislation that would legalize most of the outposts. For the first time, it would apply Israeli domestic law to the West Bank—rather than military law—a major step towards annexation. When the law passed first reading in the Knesset, one of its chief proponents said proudly: Today the Israeli Knesset moved from heading towards establishing a Palestinian state towards [Israeli] sovereignty in Judea and Samaria. Even the Israeli Attorney General has said the draft law is unconstitutional and a violation of international law. Now you may hear that the settlements are not an obstacle to peace because the settlers who don't want to leave can just stay in Palestine, like the Arab Israelis who live in Israel. But that misses a critical point: the Arab Israelis are citizens of Israel, subject to Israel's laws. Does anyone really believe the settlers will agree to submit to Palestinian law in Palestine? Likewise, some supporters of the settlements argue that the settlers could just stay in their settlements, and remain as Israeli citizens in their separate enclaves in the middle of Palestine, protected by the IDF. There are over 80 settlements east of the separation barrier, many located in places that would make a contiguous Palestinian state impossible. Does anyone seriously think that if they just stay where they are you could still have a viable Palestinian state? Some have asked, "Why can't we build in the blocs which everyone knows will eventually be part of Israel?" The reason building there or anywhere else in the West Bank now results in such pushback is that the decision of what constitutes a bloc is being made unilaterally by the Israeli government, without consultation, without the consent of the Palestinians—and without granting the Palestinians a reciprocal right to build in what will by most accounts be part of Palestine. Bottom line—without agreement or mutuality, the unilateral choices become a major point of contention. You may hear that these remote settlements aren't a problem because they only take up a small percentage of the land. Again and again we have made clear that it's not just a question of the overall amount of land available in the West Bank—it's whether the land can be connected or is broken up into small parcels that could never constitute a real state. The more outposts that are built, the more settlements expand, the less possible it is to create a contiguous state. So in the end, a settlement is not just the land it's on, it's also what the location does to the movement of people, what it does to the ability of a road to connect, what it does to the sense of statehood that is chipped away with each new construction. No one thinking seriously about peace can ignore the reality of the threat settlements pose to peace. But the problem goes well beyond just settlements: trends indicate a comprehensive effort to take West Bank land for Israel and prevent any Palestinian development there.

Today, the 60% of the West Bank known as Area C—much of which was supposed to be transferred to Palestinian control long ago under the Oslo accords—is effectively off limits to Palestinian development. Most has essentially been taken for exclusive Israeli use by unilaterally designating it as state land or including it within the jurisdiction of regional settlement councils. Israeli farms flourish in the Jordan River Valley and Israeli resorts line the shores of the Dead Sea—where Palestinian development is not allowed. In fact, almost no private Palestinian building is approved in Area C at all—only one permit was issued by Israel in all of 2014 and 2015, while approvals for hundreds of settlement units were advanced during that same period. Moreover, Palestinian structures in Area C that do not have a permit from the Israeli military are potentially subject to demolition. And they are currently being demolished at historically high rates: over 1300 Palestinians, including over 600 children, have been displaced in 2016 alone—more than any previous year. So the settler agenda is defining the future in Israel. And their stated purpose is clear: They believe in one state: greater Israel. In fact, one prominent minister who heads a pro-settler party declared just after the U.S. election that the era of the two state solution is over, and many other coalition ministers publicly reject a Palestinian state. And they are increasingly getting their way, with plans for hundreds of new units in East Jerusalem recently announced and talk of a major new settlement building effort in the West Bank to follow. So why are we so concerned? Well ask yourself these questions: What happens if they succeed? Where does that lead? There are currently about 2.75 million Palestinians living under military occupation in the West Bank, most of them in Areas A and B where they have limited autonomy. They are restricted in their daily movements by a web of checkpoints, and unable to travel into or out of the West Bank without a permit from the Israelis. So if there is only one state, you would have millions of Palestinians permanently living in segregated enclaves in the middle of the West Bank, with no real political rights, separate legal, education and transportation systems, vast income disparities, under a permanent military occupation that deprives of them of the most basic freedoms—separate but unequal. Nobody can explain how that works. Would an Israeli accept living that way? Would an American? Will the world accept it? If the occupation becomes permanent, over time the Palestinian Authority could dissolve and turn over all administrative and security responsibilities to the Israelis. What would happen then? Who would administer the schools and hospitals? Does Israel want to pay for the billions of dollars of lost international assistance that the PA now receives? Would the Israel Defense Force police the streets in every Palestinian city and town?

How would Israel respond to a growing civil rights movement from Palestinians demanding a right to vote, or widespread protests and unrest across the West Bank? How does Israel reconcile a permanent occupation with its democratic ideals? How does the U.S. continue to defend that and still live up to our own democratic ideals? Nobody has ever provided good answers to those questions because there aren't any. And there would be an increasing risk of more intense violence between Palestinians and settlers, and complete despair among Palestinians would create fertile ground for extremists. With all the external threats Israel faces, does it really want an intensifying conflict in the West Bank? How does that help Israel's security? The answer: It doesn't. Which is precisely why so many senior Israeli military and intelligence leaders—past and present—believe the two-state solution is the only real answer for Israel's long term security. One thing we do know: if Israel goes down the one state path, it will never have true peace with the rest of the Arab world, and I can say that with certainty. The Arab countries have made clear that they will not make peace with Israel without resolving the Israeli-Palestinian conflict—that's not where their loyalties or their politics are. But there is something new here. Common interests in countering Iran's destabilizing activities and fighting extremists as well as diversifying their economies have created real possibilities. I have spent a great deal of time with key Arab leaders exploring this, and there is no doubt that they are prepared to have a fundamentally different relationship with Israel. That was stated in the Arab Peace Initiative, and all my recent conversations have confirmed their readiness, in the context of Israeli-Palestinian peace, not just to normalize relations—but to work openly on securing that peace with significant regional security cooperation. Many have shown a willingness to support serious Israeli-Palestinian negotiations and to take steps on the path to normalization of relations—including public meetings—providing there is meaningful progress towards a two state solution. That is a real opportunity that should not be missed. That raises one final question: is ours the generation that gives up on the dream of a Jewish, democratic state of Israel living in peace and security with its neighbors? Because that is literally what is at stake. That is what informed our vote at the Security Council last week: the need to preserve the two state solution. And both sides must take responsibility for that.

We have repeatedly and emphatically stressed to the Palestinians that all incitement to violence must stop. We have consistently condemned all violence and terrorism. And we have strongly opposed unilateral efforts to delegitimize Israel in international fora. We have made countless public and private exhortations to the Israelis to stop the march of settlements. In literally hundreds of conversations with Prime Minister Netanyahu, I have made clear that continued settlement activity would only increase pressure for an international response. We have all known for some time that the Palestinians were intent on moving forward with a settlements resolution, and I advised the Prime Minister repeatedly that further settlement activity only invited UN action. Yet the settlement activity only increased—including advancing the unprecedented legislation to legalize settler outposts that the Prime Minister himself reportedly warned could expose Israel to action at the Security Council and even international prosecution, before deciding to support it. In the end, we could not in good conscience protect the most extreme elements of the settler movement as it tries to destroy the two state solution. We could not in good conscience turn a blind eye to Palestinian actions that fan hatred and violence. It is not in U.S. interests to help anyone on either side create a unitary state. We may not be able to stop them, but we cannot be expected to defend them. And it is certainly not the role of any country to vote against its own policies.

That is why we decided not to block the UN resolution that makes clear both sides have to take steps to save the two state solution while there is still time. We did not take this decision lightly. The Obama administration has always defended Israel against any efforts at the UN and any international fora, or biased and one-sided resolutions that seek to undermine its legitimacy or security. And that has not changed. But it's important to remember that every U.S. administration—Republican and Democratic—has opposed settlements as contrary to the prospects for peace. And action at the UN Security Council is far from unprecedented: In fact, previous U.S. Administrations of both political parties have allowed resolutions that were critical of Israel to pass, including on settlements, on dozens of occasions. Under George W. Bush alone, the Council passed six resolutions that Israel opposed—including one that endorsed a plan calling for a complete freeze on settlements, including natural growth. Let me read you the lead paragraph from a New York Times story dated December 23rd: With the United States abstaining, the Security Council adopted a resolution today strongly deploring Israel's handling of the disturbances in the occupied territories—which the resolution defined as including Jerusalem. All of the 14 other Security Council members voted in favor. That story was not written last week; it was written December 23, 1987—26 years to the day we voted last week, when Ronald Reagan was president. Yet despite growing pressure, the Obama administration held a strong line against any UN action—we were the only administration since 1967 that had not allowed any resolution to pass that.

Israel opposed. In fact, the only time in 8 years the Obama administration exercised its veto at the United Nations was against a one-sided settlements resolution in 2011 that did not mention incitement or violence. Lets look at what's happened since then: There have been over 30,000 settlement units advanced through some stage of the planning process. That's right: over thirty thousand settlement units. And if we had vetoed this resolution, the United States would have been giving license to further unfettered settlement construction that we fundamentally oppose. We reject the criticism that this vote abandons Israel. On the contrary, it is not this Resolution that is isolating Israel. It is a policy of permanent settlement construction that risks making peace impossible. Virtually every country in the world other than Israel opposes settlements. That includes many friends of Israel—including the United Kingdom, France and Russia—all of whom voted in favor of the settlements resolution in 2011 and again this year, along with every other member of the Council. In fact, this resolution simply reaffirms statements made by the Security Council on the legality of settlements over several decades; it does not break new ground. In 1978, the State Department legal advisor advised the Congress of his conclusion that the Israeli governments program of establishing civilian settlements in the occupied territory is inconsistent with international law. We see no change since then to affect that fundamental conclusion. You may have heard some criticize this resolution for calling East Jerusalem occupied territory. But to be clear, there was absolutely nothing new in last week's resolution on that issue. It was one of a long line of Security Council resolutions that included East Jerusalem as part of the territories occupied by Israel in 1967, and that includes resolutions passed by the Security Council under President Reagan and President George H. W. Bush. And remember that every U.S. administration since 1967—along with the entire international community—has recognized East Jerusalem as among the territories that Israel occupied in the Six Day War. And I want to stress this point: we fully respect Israel's profound historic and religious ties to the city and its holy sites. This resolution in no manner prejudges the outcome of permanent status negotiations on East Jerusalem, which must of course reflect those ties and realities on the ground. We also strongly reject the notion that somehow the United States was the driving force behind this resolution. The Egyptians and Palestinians had long made clear their intention to bring a resolution to a vote before the end of the year. The United States did not draft or originate this resolution, nor did we put it forward. It was drafted and ultimately introduced by Egypt, which is one of Israel's closest friends in the region, in coordination with the Palestinians and others. During the course of this process, we made clear to others, including those on the Security Council, that we would oppose any resolution that did not include language on terrorism and incitement. Making such positions clear is standard practice with resolutions at the Security Council. The Egyptians, Palestinians and many others understood that if the text were more balanced, it was possible we would not block it. But we also made crystal clear that the President would not make a final decision about our own position until we saw the final text. In the end, we did not agree with every word in this resolution. There are important issues that are not sufficiently addressed—or addressed at all. But we could not in good conscience veto a resolution that condemns violence and incitement, reiterates what has long been the overwhelming consensus international view on settlements, and calls for the parties to start taking constructive steps to advance the two state solution on the ground. Ultimately, it will be up to the Israeli people to decide whether the unusually heated attacks that Israeli officials have directed toward this administration best serve Israel's national interests and its relationship with an ally that has been steadfast in its support. Those attacks, alongside allegations of a U.S.-led conspiracy and other manufactured claims, distract and divert attention from what the substance of this vote really was about. We all understand that Israel faces very serious threats in a very tough neighborhood. Israelis are rightfully concerned about making sure that there is not a new terrorist haven next door. And Israelis are fully justified in decrying attempts to delegitimize their state and question the right of a Jewish state to exist. But this vote was not about that. It was about actions that Israelis and Palestinians are taking that are increasingly rendering a two-state solution impossible. It was not about making peace with the Palestinians now—it was about making sure peace with the Palestinians will be possible in the future. Now we all understand that Israel faces very serious threats in a very tough neighborhood. And Israelis are rightfully concerned about making sure that there is not a new terrorist haven next door. But this vote is not about making peace with the Palestinians now—it is about making sure peace with the Palestinians is possible in the future. So how do we get there? Since the parties have not yet been able to resume talks, the U.S. and the Middle East Quartet have repeatedly called on both sides to independently demonstrate a genuine commitment to the two state solution—not just with words, but with real actions and policies—to create the conditions for meaningful negotiations. We have called for both sides need to take significant steps on the ground to reverse current trends and send a clear message that they are prepared to fundamentally change the equation—without waiting for the other side to act. We have pushed them to comply with their basic commitments under their own prior agreements in order to advance a two state reality on the ground. We have called for the Palestinians to do everything in their power to stop violence and incitement, including publicly and consistently condemning acts of terrorism and stopping the glorification of violence.

We have called on them to continue efforts to strengthen institutions and improve governance. And we have stressed that the Hamas arms build-up and militant activities in Gaza must stop. Along with our Quartet partners, we have called on Israel to end the policy of settlement construction and expansion, taking of land for exclusive Israeli use, and denying Palestinian development. To reverse this process, the US and our partners have encouraged Israel to resume the transfer of civil authority to the Palestinians in Area C, consistent with the transition called for by Oslo. We have made clear that significant progress across a range of sectors —including housing, agriculture, and natural resources—can be made without negatively impacting Israel's legitimate security needs. And we've called for significantly easing the movement and access restrictions to and from Gaza, with due consideration for Israel's need to protect its citizens from terrorist attacks. Let me stress here again: none of these steps would negatively impact Israel's security. Let me also emphasize—this is not about offering limited economic measures that perpetuate the status quo. We are talking about significant steps that would signal real progress towards creating two states. That's the bottom line: If we are serious about the two state solution, it is time to start implementing it now. Advancing the process of separation now, in serious way, could make a significant difference in saving the two state solution. And much progress can be made in advance of negotiations, as contemplated by the Oslo process—in fact, these steps will help create the conditions for successful talks. In the end, we all understand that a final status agreement can only be achieved through direct negotiations between the parties, because peace cannot be imposed. There are other countries in the UN who believe it is our job to dictate the terms of a solution in the Security Council. Others want us to simply recognize a Palestinian state absent an agreement. These are not the choices we will make. We choose instead, drawing on the experiences of the past eight years, to provide a way forward when the parties are ready for serious negotiations. In a place where the narratives from the past powerfully inform the present, it's important to understand the history. We mark this year and next a series of milestones that I believe both illustrate the two sides of the conflict and form the basis for its resolution. It's worth touching on them briefly. 120 years ago, the First Zionist Congress was convened in Basel by a group of Jewish visionaries who decided that the only effective response to the waves of anti-Semitic horrors sweeping across Europe was to create a state in the historic home of the Jewish people, where their ties to the land went back centuries—a state that could defend its borders, protect its people, and live in peace with its neighbors. That was the modern beginning and it remains the dream of Israel today. Nearly seventy years ago, United Nations General Assembly Resolution 181 finally paved the way to making the State of Israel a reality. The concept was simple: create two states for two peoples, one Jewish and one Arab, to realize the national aspirations of both Jews and Palestinians. Both Israel and the PLO referenced Resolution 181 in their respective Declarations of Independence. The United States recognized Israel seven minutes after its creation—but the Palestinians and the Arab world did not, and from its birth Israel had to fight for its life. Palestinians also suffered terribly in that 1948 war, including many who had lived for generations in a land that had long been their home too. When Israel celebrates its 70th anniversary in 2018, the Palestinians will mark a very different anniversary: 70 years since what they call the *Nakba*, or catastrophe. Next year will also mark 50 years since the end of the Six-Day War, when Israel again fought for its survival. And Palestinians will again mark just the opposite: 50 years of military occupation. Both sides have accepted UN Security Council Resolution 242, which called for the withdrawal of Israel from territory it occupied in 1967 in return for peace and secure borders, as the basis for ending the conflict. It has been more than twenty years since Israel and the PLO signed their first agreement—the Oslo Accords—and the PLO formally recognized

Israel. Both sides committed to a plan to transition much of the West Bank and Gaza to Palestinian control during permanent status negotiations that would put an end to their conflict. Unfortunately, neither the transition nor the final agreement came about—and both sides bear responsibility for that. Finally, some 15 years ago King Abdullah of Saudi Arabia came out with the historic Arab Peace Initiative, which offered fully normalized relations with Israel when it made peace—an enormous opportunity then and now, which has never fully been embraced. That history was critical to our approach to trying to resolve the conflict. And based on my experience with both sides over the last four years, including the nine months of formal negotiations, the core issues can be resolved if there is leadership on both sides committed to finding a solution. In the end, I believe the negotiations did not fail because the gaps were too wide—but because the level of trust was too low. Both sides were concerned that any concessions would not be reciprocated—and would come at too great a political cost. And the deep public skepticism only made it more difficult to take risks. In the countless hours we spent working on a detailed framework, we worked through numerous formulations and developed specific bridging proposals. And we came away with a clear understanding of the fundamental needs of both sides. In the past two and a half years, I have tested ideas with key regional and international stakeholders, including our Quartet partners. I believe what has emerged is a broad consensus on balanced principles that would satisfy the core needs of both sides.

President Clinton deserves great credit for laying out extensive parameters designed to bridge gaps in advanced final status negotiations sixteen years ago. Today, with mistrust too high to even start talks, we are at the opposite end of the spectrum. Neither side is willing to even risk acknowledging the others bottom line, and more negotiations that do not produce progress will only reinforce the worst fears. Everyone understands that negotiations would be complex and difficult, and nobody can be expected to agree on the final result in advance. But if the parties could at least demonstrate that they understand the other side's most basic needs—and are potentially willing to meet them if theirs are also met at the end of comprehensive negotiations—enough trust could be established to enable a meaningful process to begin. It is in that spirit that we offer the following principles: not to prejudge or impose an outcome, but to provide a possible basis for serious negotiations when the parties are ready. Individual countries may have more detailed policies on these issues—as we do—but I believe there is broad consensus that a final status agreement that could meet the needs of both sides would:

1. Provide for secure and recognized international borders between Israel and a viable and contiguous Palestine, negotiated based on the 1967 lines with mutually agreed equivalent swaps. Resolution 242, which has been enshrined in international law for 50 years, provides for the withdrawal of Israel from territory it occupied in 1967 in return for peace with its neighbors and secure and recognized borders. It has long been accepted by both sides, and it remains the basis for an agreement today. As Secretary, one of the first issues I worked out with the Arab League was their agreement that the reference in the Arab Peace

Initiative to the 1967 lines included the concept of land swaps, which the Palestinians have acknowledged. This is necessary to reflect practical realities on the ground, and mutually agreed equivalent swaps will ensure the agreement is fair to both sides. There is also broad recognition of Israel's need to ensure that the borders are secure and defensible, and that the territory of Palestine is viable and contiguous. Virtually everyone I have spoken to has been clear on this principle as well: No changes by Israel to the 1967 lines will be recognized by the international community unless agreed to by both sides.

2. Fulfill the vision of UN General Assembly Resolution 181 of two states for two peoples, one Jewish and one Arab, with mutual recognition and full equal rights for all their respective citizens. This has been the foundational principle of the two state solution from the beginning: Creating a state for the Jewish people and a state for the Palestinian people, where each can achieve their national aspirations. And resolution 181 is incorporated into the foundational documents of both the Israelis and Palestinians. Recognition of Israel as a Jewish state has been the U.S. position for years, and based on my conversations, I am convinced many others are now are prepared to accept it as well—provided the need for a Palestinian state is also addressed.
   We also know there are some 1.7 million Arab citizens who call Israel their home and must now and always be able to live as equal citizens, which makes this a difficult issue for Palestinians and others in the Arab world. That is why it is so important that in recognizing each other's homeland—Israel for the Jewish people and Palestine for the Palestinian people—both sides reaffirm their commitment to upholding full equal rights for all of their respective citizens.
3. Provide for a just, agreed, fair and realistic solution to the Palestinian refugee issue, with international assistance, that includes compensation, options and assistance in finding permanent homes, acknowledgment of suffering and other measures necessary for a comprehensive resolution consistent with two states for two peoples. The plight of many Palestinian refugees is heartbreaking, and all agree their needs must be addressed. As part of a comprehensive resolution, they must be provided with compensation, their suffering must be acknowledged, and there will need to be options and assistance in finding permanent homes. The international community can provide significant support and assistance, including in raising money to help ensure the compensation and other needs of the refugees are met, and many have expressed a willingness to contribute. But there is general recognition that the solution must be consistent with two states for two peoples, and cannot affect the fundamental character of Israel.
4. Provide an agreed resolution for Jerusalem as the internationally recognized capital of the two states, and protect and assure freedom of access to the holy sites consistent with the established *status quo*. Jerusalem is the most sensitive issue for both sides, and the solution must meet the needs not only of the parties, but of all three monotheistic faiths. That is why the holy sites that are sacred to billions of people around the world must be protected and remain accessible,

and the established status quo maintained. Most acknowledge that Jerusalem should not be divided again like it was in 1967. At the same time, there is broad recognition that there will be no peace agreement without reconciling the basic aspirations of both sides to have capitals there.

5. Satisfy Israel's security needs and bring a full end to the occupation, while ensuring that Israel can defend itself effectively and that Palestine can provide security for its people in a sovereign and non-militarized state. Security is the fundamental issue for Israel. Everyone understands that no Israeli government can ever accept an agreement that does not satisfy its security needs or risks creating an enduring security threat like Gaza in the West Bank. And Israel must be able to defend itself effectively, including against terrorism and other regional threats. In fact, there is a real willingness by Egypt, Jordan and others to work together with Israel on meeting key security challenges. I believe these collective efforts, including close coordination on border security, intelligence sharing, and joint operations, can play a critical role in securing the peace. At the same time, fully ending the occupation is the fundamental issue for the Palestinians: They need to know that the military occupation will really end after an agreed transitional process, and that they can live in freedom and dignity in a sovereign state while providing security for their population even without a military of their own. This is widely accepted as well.
   Balancing those requirements was among the most important challenge we faced in the negotiations, but one where the United States could provide the most assistance. That's why a team led by General John Allen, one of our nation's foremost military minds, and dozens of experts from the Department of Defense and other agencies, engaged extensively with the IDF on trying to find solutions that could help Israel address its legitimate security needs. They developed innovative approaches to creating unprecedented, multi-layered border security, enhancing Palestinian capacity, and enabling Israel to retain the ability to address threats by itself even when the occupation had ended. General Allen and his team were not suggesting any particular outcome or timeline—they were simply working on ways to support whatever the negotiators agreed to. And they did some very impressive work that gives me confidence that Israel's security requirements can be met.
6. End the conflict and all outstanding claims, enabling normalized relations and enhanced regional security for all as envisaged by the Arab Peace Initiative. It is essential for both sides that the final status agreement resolves all the outstanding issues and finally brings closure to the conflict, so they can move ahead to a new era of peaceful coexistence and cooperation. For Israel, this must also bring broader peace with its Arab neighbors. That is the fundamental promise of the Arab Peace Initiative, which key Arab leaders have affirmed. The API also envisions enhanced security for all in the region. This is the area where Israel and the Arab world are looking at the greatest moment of potential transformation in the Middle East since Israel's creation in 1948. The Arab world faces its own set of security challenges. With Israeli-Palestinian peace, Israel, the

United States, Jordan and Egypt—together with the GCC countries—would be ready and willing to define a new security partnership for the region that would be groundbreaking. Ladies and Gentlemen: it's vital that we all work to keep open the possibility of peace, and that we not lose hope in the two state solution, no matter how difficult it may seem—because there really is no viable alternative. We all know that a speech alone won't produce peace. But based on over 30 years of experience and the lessons from the past 4 years, I have suggested a practical path that the parties could take. Realistic steps on the ground now, consistent with the parties own prior commitments, that will begin the process of separating into two states, a political horizon to work towards to create the conditions for successful final status talks, and a basis for negotiations that the parties could accept to demonstrate that they are serious about making peace. We can only encourage them to take this path, we cannot walk down it for them. But if they ever take those steps, peace would bring extraordinary benefits in enhancing the security, stability and prosperity of Israelis, Palestinians and the entire region. The Palestinian economy has amazing potential in the context of independence, with major private sector investment possibilities and a talented young workforce. And Israel's economy could enjoy unprecedented growth as it becomes a regional economic powerhouse, taking advantage of its unparalleled culture of innovation and trading opportunities with new Arab partners. Meanwhile, security challenges could be addressed by an entirely new security arrangement, in which Israel cooperates openly with key Arab states. That is the future that everyone should be working for. President Obama and I know that the incoming Administration has signaled that they will take a different path, and even suggested breaking from long-standing U.S. policies on settlements, Jerusalem—and possibly the two state solution. That is for them to decide. But we cannot—in good conscience—do nothing, and say nothing, when we see the hope of peace slipping away. This is a time to stand up for what is right. We have long known what two-states, living side by side, in peace and security looks like. We should not be afraid to say so. I really began to reflect on what we have learned—and the way ahead—when I recently joined President Obama in Jerusalem for the state funeral for Shimon Peres. Shimon was one of the founding fathers of Israel who became one of the world's great elder-statesmen. I was proud to call him my friend, and I know President Obama was as well. I remembered the first time I saw Shimon in person—standing on the White House lawn for the signing the historic Oslo Accords. And I thought about the last time, at an intimate Shabbat dinner just a few months before he died when we toasted to the future of Israel and to the peace he still so passionately believed in for his people. He summed it up simply and eloquently, as only Shimon could: The original mandate gave the Palestinians 48%, now it's down to 22%. I think 78% is enough for us. As we laid Shimon to rest that day, many of us couldn't help but wonder if peace between Israelis and Palestinians was also being buried along with one of its most eloquent champions. We cannot let that happen, that there was simply too much at stake—for future generations of Israelis and Palestinians—to give into pessimism, especially when peace is in fact still possible. We must not lose hope in the possibility of peace. We must not give into those who say what is now must always be, that there is no chance for a better future. Ultimately, it is up to Israelis and Palestinians to make the difficult choices for peace—and if they are, we can all help. And for the sake of future generations of Israelis and Palestinians, for all the people of the region, and for the United States, let's hope they are prepared to make those choices now. Thank you."

Source: Haaretz, 2016.

Bibliography

Abukhater, Ahmed. (2009). Through the Eyes of the Indigenous Population: Innovative and Comprehensive Approaches to Mediation and ADR in the Middle East, *Peace Conflict & Development Journal*, June, issue 14.

Abukhater, Ahmed. (2010). On the Cusp of Water War: a diagnostic account of the volatile geopolitics of the Middle East, *Peace and Conflict Studies Journal*, 17(2).

Abukhater, Ahmed. (2013). *Water as a Catalyst for Peace, Transboundary Water Management and Conflict Resolution*. Routledge.

Agha, Hussein and Robert Malley. (2002). The Last Negotiation: How to End the Middle East Peace Process, *Foreign Affairs*, Vol. 81, No. 3.

Aljazeera. (2003). *The history of Palestinian revolts*. Retrieved from: https://www.aljazeera.com/archive/2003/12/2008410112850675832.html.

Aljazeera. (2016). *UN Security Council urges end to Israeli settlements*. Retrieved from: http://www.aljazeera.com/news/2016/12/passes-resolution-israeli-settlements-161223192709807.html.

Allouche, J. (2005). *Water Nationalism: an Explanation of the Past and Present Conflicts in Central Asia, the Middle East, and the Indian Subcontinent?* Universite de Geneve, Ph.D. Dissertation.

Alatout, S. (2003). *Imagining hydrological boundaries, constructing the Nation State: a 'fluid' history of Israel, 1936–1959.* Ph.D. Dissertation, Cornell University.

Alatout, S. (2006). Towards a bio-territorial conception of power: Territory, population, and environmental narratives in Palestine and Israel, *Political Geography,* 25, pp. 60–621.

Amnesty International. (2009). *Troubled Waters – Palestinians Denied Fair Access to Water.*

Ashrawi, Hanna. Interview with Matthew Amroliwala. *BBC World News*. 22 January, 2018. Retrieved from https://www.youtube.com/watch?v=luBD8Akh7V4.

Ashrawi, Hanna. Interview with Matthew Amroliwala. *BBC World News*. 6 December, 2017. Retrieved from https://www.youtube.com/watch?v=2bdil4YadxE.

Aslan Reza. Interview with The Stream on Trump's Jerusalem move. *Al Jazeera*. 7 December, 2017.

BBC News. (1999). World: *Middle East - Drought 'forces Israel to break treaty.'* Retrieved April 10th, 20017 from: www.news.bbc.co.uk/2/hi/middle_east/296797.stm.

Blumenthal, Max. (2015). *The 51 Day War: Ruin and Resistance in Gaza.* Nation Books.

Brooks D. B. and Mehmet, O. (2000). *Water Balances in the Eastern Mediterranean.* The International Development Research Centre.

Burgess, H. and Burgess, G. (2007). Constructive confrontation: A transformative approach to intractable conflicts, *Conflict Resolution Quarterly*, 13(4), pp. 305–322.

A. Abukhater, *Palestine - Peace by Piece*,
https://doi.org/10.1007/978-3-030-03985-1

Caponera, D. A. (1985). Patterns of Cooperation in international water law: Principles and institutions. *Natural Resources Journal* 25, July: pp. 563–588.

Cartoon Movement. (2017). *Cartoon showing the impact of moving the US embassy to Jerusalem on the Middle East Peace (Throwing a Grenade at the Dove of Peace).* Retrieved from https://www.cartoonmovement.com/cartoon/44787.

Chris Hedges. Speech. *Revolution Books Town Hall Meeting at Ethical Culture Society.* January 13, 2009. Retrieved from: https://www.youtube.com/watch?v=ZTRGxgzjVtI.

Christison, Kathleen. (1999). *Perceptions of Palestine: Their Influence on U.S. Middle East Policy.* Berkeley: University of California Press.

Clemens, Michaela S. (2007). The Influence of Refugee Status on Palestinian Identity and the Impact of Identity on Durable Solutions to the Refugee Problem. *Nebraska Anthropologist.* 28.

CNN Library. (2017). Yasser Arafat Fast Facts, *CNN*, Mon June 19. Retrieved from http://www.cnn.com/2013/09/10/world/meast/yasser-arafat-fast-facts/.

Cody, Edward. (1988). PLO-Israeli Talks Asked by Arafat. *The Washington Post*, December 14, 1988. Retrieved from https://www.washingtonpost.com/archive/politics/1988/12/14/plo-israeli-talks-asked-by-arafat/b491d33f-43ed-4612-a872-d7be40ffbf10/?utm_term=.5e0d109cd7f7.

Corradin, Camilla. (2016). Israel deliberately denies Palestinians control over their water sources and sets the ground for water domination, AL Jazeera. Retrieved from: https://www.aljazeera.com/news/2016/06/israel-water-tool-dominate-palestinians-160619062531348.html.

Dajani, Omar. (2007). Shadow or Shade? The Roles of International Law in Palestinian-Israeli Peace Talks, *Yale Journal of International Law,* Volume 32, issue 1.

Daoudy, Marwa (Lecturer, Department of Political Science, the Graduate Institute of International Studies, Geneva), personal phone interview by author, Austin, TX, January 29, 2009.

Dellapenna, Joseph W. (2002). Markets-Ethics-Law: what can each contribute? *Conference Proceedings: From Conflict to Co-operation in International Water Resources Management: Challenges and Opportunities, International Conference,* 20–22 November 2002, UNESCO-IHE Institute for Water Education, Delft, Netherlands.

Detriot News. (1979). *Cartoon* retrieved from http://www.original-political-cartoon.com/cartoon-gallery/buy/thanksgiving-parade/2851/.

Diabes-Murad, Fadia. (2004). *Water Resources in Palestine: A Fact Sheet and Basic Analysis of Legal Status.* International Water Law Research Institute, Dundee University, Scotland.

Dolatyar, M., and Gray, T. S. (2000). *Water Politics in the Middles East.* Macmillan Press LTD.

Elbedour, Salman, David T. Bastien and Bruce A. Center. (1997). Identity Formation in the Shadow of Conflict: Projective Drawings by Palestinian and Israeli Arab Children from the West Bank and Gaza. *Journal of Peace Research*, 34(2):217–231.

Elmusa, S. (1995). The Jordan-Israel Water Agreement: A Model or an Exception? *Journal of Palestine Studies,* 24(3), pp. 63–73.

Encyclopedia Britannica Online. (2010). [*Map*]. Retrieved April 13, 2018 from https://www.britannica.com/event/Peel-Commission?oasmId=147261.

Erakat, Nour. (2014). Five Israeli Talking Points on Gaza—Debunked, *The Nation*. Retrieved from https://www.thenation.com/article/five-israeli-talking-points-gaza-debunked/.

Fanack after UNEP. (2002). *Water Resources*. Retrieved from https://water.fanack.com/palestine/water-resources/.

FAO- Aquastat. (2002). *Water availability per person per year.*

Faruqui, N. I. (2000). *Water Management in Islam.* United Nations University.

Farsoun, Samih. (1998). *Palestine and the Palestinians*. Routledge.

Fischhendler, Itay. (2008). Ambiguity in Transboundary Environmental Dispute Resolution: The Israeli–Jordanian Water Agreement, *Journal of Peace Research*, 45(1), pp. 79–97.

Frisch, Hillel. (2000). *Water and Israel's National Security.* The Begin-Sadat (BESA) Center for Strategic Studies, Bar Ilan University, Israel.

George Orwell (1946). *Politics and the English Language*. Penguin Group.

Giordano, M. and A. Wolf. (2001). Incorporating Equity into International Water Agreements, *Social Justice Research*, 14(4): 349–366.

Gilad, Elon. (2015). What Israelis Call Palestinians and Why It Matters. Haaretz. Retrieved from https://www.haaretz.com/israel-news/.premium-what-israelis-call-palestinians-and-why-it-matters-1.5424214.

Golan. (2007). *Water Issues of the Golan Heights* [Electronic Version]. Retrieved 08/13/07 from: http://www.golan.org.il/water.html.

Greenberg, J. (2006). River Jordan: Sacred, tainted and dying, *Chicago Tribune*.

Gleick, Peter. (1994). Water, War, and Peace, in the Middle East, *Environment*, 36(3).

Green Olive Tours. (2014). *Cartoon by Mike Vukovich depicting the "Two-State Solution" in reality.* Retrieved from http://blog.toursinenglish.com/2014/09/gaza-context.html.

Haddadin, Munther J. (2002). *Diplomacy on the Jordan: International Conflict and Negotiated Resolution.* Norwell, MA: Kluwer Academic Publishers.

Haddadin, Munther J. (2002). Water in the Middle East peace process, *The Geographical Journal*, 168 (4), pp 324–340.

Hass, Amira. (2003). You Can Drive Along and Never See an Arab, *Haaretz*. Retrieved from https://www.haaretz.com/1.4810272.

Haaretz. (2016). Full Transcript: Kerry Blasts Israeli Government, Presents Six Points of Future Peace Deal, *Haaretz*, Dec 28, 2016. Retrieved from https://www.haaretz.com/israel-news/criticizing-netanyahu-warning-against-one-state-kerry-s-full-speech-1.5479380.

Hammond, Jeremy. (2017). The New York Times on "Nakba": Whitewashing the Ethnic Cleansing of Palestine, *Foreign Policy Journal.* Retrieved November 23, 2018 form https://www.foreignpolicyjournal.com/2017/05/15/the-new-york-times-on-nakba-whitewashing-the-ethnic-cleansing-of-palestine/.

Hazony, David. (2015). How Israel Is Solving the Global Water Crisis, *The Tower*, Issue 31. Accessed 1/2/2017 from http://www.thetower.org/article/how-israel-is-solving-the-global-water-crisis/.

International Law Association. (1967). *The Helsinki Rules on the Uses of the Waters of International Rivers. Adopted by the International Law Association at the fifty-second conference, held at Helsinki in August 1966. Report of the Committee on the Uses of the Waters of International Rivers,* London.

International Law Association. (2003). *Helsinki Revision: Sources of the International Law Association Rules on Water Resources.*

Irani, G. E. (2000). Islamic Mediation Techniques for Middle East Conflicts, Mediate [Electronic Version]. Retrieved 10/02/2016 from: http://www.mediate.com/articles/mideast.cfm.

Islam, S. and Susskind L. E. (2013). *Water Diplomacy: A Negotiated Approach to Managing Complex Water Networks.* Routledge.

Israel: Water as a tool to dominate Palestinians.

Jarvis, Todd and Wolf, A. (2010). Managing Water Negotiations and Conflicts in Concepts and in Practice. *In Earle, Anton, Jägerskog, A. and Öjendal, J. (ed.), Transboundary Water Management: Principles and Practice.* Earthscan.

Jerusalem Media and Communications Centre (JMCC). (2009). *Hamas-Fateh conflict.* Retrieved from http://www.jmcc.org/fastfactspag.aspx?tname=13.

Kally, Elisha. (1993). Water and Peace: *Water Resources and the Arab-Israeli Peace Process.* Tel Aviv University.

Kerr, Malcolm. (1975). *The Elusive peace in the Middle East State.* University of New York Press.

Khalidi, Rashid. (2013). *Brokers of Deceit: How the U.S. Has Undermined Peace in the Middle East.* Beacon Press.

Khalidi, Walid [Ed.]. (1991). *All That Remains: The Palestinian Villages Occupied and Depopulated by Israel in 1948.* Berkeley: University of California Press.

Khalel, Sheren. (2016). Israel's water cuts: West Bank 'in full crisis mode,' *Aljazeera*. Retrieved from: https://www.aljazeera.com/news/2016/06/israel-water-cuts-west-bank-full-crisis-mode-160620133005799.html.

Khoury, Jack. (2018). Palestinians Respond to Trump: If Jerusalem Off Table, Then U.S. No Longer Has Seat at the Table. *Haaretz*. Retrieved from https://www.haaretz.com/israel-news/palestinians-to-trump-if-jerusalem-off-table-then-so-is-u-s-1.5766591.

Klieman, A. S. (2000). *Compromising Palestine: a guide to final status negotiations.* Tel-Aviv University.

Korf, B. and H. Funfgeld. (2005). War and the commons: Assessing the changing politics of violence, access and entitlements in Sri Lanka, *Geoforum*, 37, pp. 391–403.

Le Pelley. (1977). The Christian Science Monitor, September 19, 1977. Retrieved from http://mideastcartoonhistory.com/1977to1988/1977to1988.html.

Le Pelley, The Christian Science Monitor, January 11, 1978. Retrieved from http://mideastcartoonhistory.com/1977to1988/1978.html.

Library of the Congress (2018). *Edmund Valtman, the Cartoonist Who Came in From the Cold.* Retrieved from https://www.loc.gov/rr/print/swann/valtman/presentation.html.

Lillian Goldman Law Library. (2008). *The ISRAEL-JORDAN COMMON AGENDA*, September 14, 1993, Washington, D.C. Retrieved January 24, 2009, from: http://avalon.law.yale.edu/20th_century/pal07.asp.

Linskey, M. and Heifitz, R. (2002). *Leadership on the line*. Cambridge, MA: Harvard Business School.

Lonergan, Stephen C. and David B. Brooks. (1993). *Economic, Ecological, and Geopolitical Dimensions of Water In Israel*. The Centre for Sustainable Regional Development (CSRD), the University of Victoria.

Louka Elli. (2006). *International environmental law: fairness, effectiveness, and world order.* Cambridge University Press.

Lowi, M. R. (1995). *Water and Power: the Politics of a Scarce Resource in the Jordan River Valley.* Cambridge University Press.

Lustick, Ian. (1994). *For the Land of the Lord.* Council on Foreign Relations, New York.

Malhotra, Deepak and Max H. Bazerman. (2008). *Negotiation Genius: How to Overcome Obstacles and Achieve Brilliant Results at the Bargaining Table and Beyond.* Bantam.

McDermott, Thomas. (2009). Water, Economics and the Jordan Valley, *Transboundary Water Crises conference*, New Mexico State University, January 22, 2009.

Mitchell, Christopher R. (2006). Conflict, Social Change and Conflict Resolution. An Enquiry. *In Social Change and Conflict Transformation.* Berghof Research Center for Constructive Conflict Management.

Mondoweiss. (2018). *Cartoon by Latuff about Mike Pence's visit to Israel.* Retrieved from http://mondoweiss.net/2018/01/administration-palestinians-jerusalem/.

Mnookin, Robert and Lewis Kornhauser. (1979). Bargaining in the Shadow of the Law: The Case of Divorce, *The Yale Law Journal*, Vol. 88, No. 5, Dispute Resolution, pp. 950–997.

Munayyer, Yousef. (2017). Reframing the 1967 War, *the New Yorker.* Retrieved from https://www.newyorker.com/news/news-desk/reframing-the-1967-war.

Nabulsi, Karma. (2006). *The great catastrophe, The Guardian.* Retrieved April 13, 2018 from https://www.theguardian.com/world/2006/may/12/israel1.

Naff, T. and Matson, R. C. (1984). *Water in the Middle East, Conflict or Cooperation?* Boulder, Colo., Westview.

Nile TV International. (2016). *Cartoon by Latuff, showing Israel's policies - war in Gaza and settlement building the West Bank.* Retrieved from http://www.nileinternational.net/en/?p=43097.

Noura Erakat. Interview with *MSNBC*. December 12, 2017.

Online Vijesti. (2018). *Image of the Israeli Wall.* Retrieved from http://www.vijesti.me/svijet/izrael-nastavlja-da-ilegalno-naseljava-palestinsku-zemlju-103409/galerija?p=2.

Palestinian Academic Society for the Study of International Affairs (PASSIA). (2002). *Water: The Blue Gold of the Middle East*. Palestine, Jerusalem.

Palestinian Academic Society for the Study of International Affairs (PASSIA). (2018). *UN Partition Plan 1947 and land seized by Zionist militants 1949.* [Map]. Retrieved March 12, 2018, from Passia http://palmissionuk.org/discovering+palestine/geography/un+partition+and+land+seized+by+zionist+terror+groups/.

Palestinian National Information Center. (1999). *Dangers Threatening Water Resources in Palestine.* States Information Service, PNA.

Palestine News Network (PNN). (2017). *Palestine FM responds to Theresa May's pride in Balfour Declaration.* Retrieved May, 2018 from http://english.pnn.ps/2017/10/26/palestine-fm-responds-to-theresa-mays-pride-in-balfour-declaration/.

Pappe, Ilan. (2007). *The Ethnic Cleansing of Palestine.* Oneworld Publications.

Peet. R and Watts, M. (2004). *Liberation Ecologies: Environment, Development, Social Movements.* Routledge, London, UK.

Peteet, Julie. (1997). Icons and Militants: Mothering in the Danger Zone. *Signs* 23(1):103–129.

Phillips, David (Freelance Consultant, England), personal phone interview by author, Austin, TX, December 26, 2008.

Postel, S. (1999). *Pillar of Sand: Can the Irrigation Miracle Last?* New York: W. W. Norton & Company.

Postel, S. and A. Wolf. (2001). Dehydrating Conflict, *Foreign Policy*. pp. 60–67.

Ramallah News. (2018). [*Photo*] retrieved from: http://24.ae/article.aspx?ArticleId=433336.

Reinhart, T. (2006). *The Road Map to Nowhere: Israel/Palestine since 2003.* Verso Books.

Rouyer, Alwyn R. (2000). *Turning Water into Politics: The Water Issue in the Palestinian-Israeli Conflict.* Department of Political Science, University of Idaho.

Said, Edward W. and Hitchens, Christopher. (2001). *Blaming the Victims: Spurious Scholarship and the Palestinian Question.* Verso.

Said, Edward. (2003). Memory, Inequality, and Power: Palestine and the Universality of Human Rights, *UC Berkeley Educational Technology Services*. Retrieved from https://www.youtube.com/watch?v=Pb2pYStv8x8.

Scheumann, W. (1998). *Water in the Middle East: Potential for Conflicts and Prospects for Cooperation.* Technical University of Berlin.

Schwartz, K. (2006). "Masters in Our Native Place": The politics of Latvian national parks on the road from Communism to "Europe," *Political Geography*, 25, pp. 42–71.

Shamir, U., and Haddadin, M. (2003). *Jordan Case study: THE Jordan River Basin, Part I and II.* UNESCO-IHP, PCCP Series Publication.

Shmueli, D.F. (1999). Approaches to Water Dispute Resolution: Applications to Arab-Israeli Negotiations, *International Negotiation*, 4(2), pp. 295–325.

Söderblom, Jason D. (2003). *State of Inequity: The UN Partition Plan of 1947.* The Terrorism Intelligence Centre, Canberra.

Sosland, Jeffrey. (2007). *Cooperating Rivals: The Riparian Politics of the Jordan River Basin.* SUNY Press.

SPIDR. (1997). *Best Practices for Government Agencies: Guidelines for Using Collaborative Agreement-Seeking Processes.*

Spiegel, Carolin. (2005). *International Water Law: the Contributions of Western United States Water Law to the United Nations Convention on the Law of the Non-navigable Uses of International Watercourses.* Duke University School of Law.

Susiman, D., *et al.* (2005). *The Arc - A Formal Structure for a Palestinian State.* RAND Corporation.

Tamimi, A. (2005). *Palestine Roots of Conflict (presentation).* Loughborough.

Tacsan, Joaquin. (1991). *The Dynamics of International Law in Conflict Resolution.* Springer.

TEDx Boulder. (2016, September 17th). *Ahmed Abukhater: Water for Peace: extinguishing the fire of war and conflict.* [Video file]. Retrieved May 14, 2017, from https://www.youtube.com/watch?v=s3yVCoAwreM.

The Britain Israel Communications & Research Centre, BICOM. (2007). *Israel Maps – the Golan Heights* [Electronic Version]. Retrieved, 08/13/07 from: http://www.bicom.org.uk/cms_image/?file_id=39&w=430&h=600&c=1.

The Daily Mail. (1977). Cartoon by Mac, November 21. Retrieved from http://mideastcartoonhistory.com/1977to1988/1977to1988.html.

The Guardian. (2014). From the archive: 6 September 1972: *Munich massacre captured on screen.* Retrieved from https://www.theguardian.com/theguardian/2014/sep/06/archive-1972-munich-massacre-on-television.

The Israel-Palestinian Interim Agreement on the West Bank and the Gaza Strip. (1995). Retrieved from: http://www.mfa.gov.il/mfa/foreignpolicy/peace/guide/pages/the%20israeli-palestinian%20interim%20agreement.aspx.

The Untied Nation Department of Public Information (UNDPI). (2003). *The Question of Palestine and the United Nations.* United Nations Department of Public Information. Retrieved December 26, 2017, from http://www.un.org/Depts/dpi/palestine/ch2.pdf.

Transboundary Freshwater Dispute Database (TFDD) (2008). Oregon State University. [Electronic Version]. Retrieved 04/15/2017 from: http://www.transboundarywaters.orst.edu/.

Trip'ol'II. (2007). *Occupation 101 [documentary].* Retrieved 08/15/17 from: www.occupation101.com.

United Nations Environment Programme (UNEP). (2002). *Global Environment Outlook 3: Past, Present and Future Perspectives.* Earthscan, London.

Virtual Jewish Library. (1996). *1996 Likud Party Platform* [Electronic Version]. Retrieved 08/13/2017 from: http://www.jewishvirtuallibrary.org/jsource/Politics/likud.html.

Visualizing Palestine. (2013). *Palestinian-Israeli Peace Talks and Expanding Settlements.* [Infographic]. Retrieved March 23, 2018, from https://visualizingpalestine.org/visuals/palestinian-israeli-peace-talks-settlements-oslo.

Visualizing Palestine. (2018). *Palestine Shrinking Expanding Israel.* [Map]. Retrieved March 23, 2018, from https://visualizingpalestine.org/visuals/shrinking-palestine?v=large.

Wagner, Donald E. and Davis, Walter T. (2014). *Zionism and the Quest for Justice in the Holy Land.* Pickwick Publications.

Westing, A. H. (1986). Environmental factors in strategic policy and action: an overview. In: A.H. Westing (ed.), *Global Resources and International Conflict: Environmental Factors in Strategic Policy and Action.* New York: Oxford University Press.

Whiteman, Marjorie M. (1964). *Digest of International Law,* Vol. 3. Washington DC: Department of State.

Wolf, A. (1995). Criteria for Equitable Allocations: The heart of international water conflict, *Natural Resources Forum.* Vol. 23 #1, February 1999. pp. 3–30.

Wolf, A. (1996). *Middle East Water Conflicts and Directions for Conflict Resolution.* International Food Policy Research Institute, Washington, D.C.

Wolf, A. (1998). Conflict and Cooperation Along International Waterways, *Water Policy,* 1 (2): 251–265.

Wolf, A. (1999). Criteria for Equitable Allocations: The heart of international water conflict, *Natural Resources Forum.* Vol. 23 #1, February 1999. pp. 3–30.

Wolf, Aaron T and Newton, Joshua T. (2008). Case Studies of Transboundary Dispute Resolution. *Appendix C in, Delli Priscoli, Jerry and Aaron T. Wolf. Managing and Transforming Water Conflicts.* Cambridge: Cambridge University Press.

Woodhouse, Melvin (implementation advisor on the ORACOM and the SADC Protocol at an operational level), personal phone interview by author, Austin, TX, December 13, 2008.

Wright, Robin. (2018). Why the Palestinians Are Boycotting the Trump Administration. *The New Yorker.* Retrieved from https://www.newyorker.com/news/news-desk/the-man-in-the-middle-of-trump-palestinian-tensions.

WWF. (2016). *Water Risk Map.* Retrieved August 12, 2016, from http://wwf.panda.org/what_we_do/how_we_work/our_global_goals/water/freshwater_news/?295473/WWF-and-AWS-join-forces-to-strengthen-water-stewardship.

Zawahri, Neda A. (2004). *The Water Weapon: Havoc and Harmony over International Rivers.* Department of Politics, University of Virginia.

Zawahri, Neda. (2009). Attempts at Governing the Jordan River System, *Transboundary Water Crises conference,* New Mexico State University, January 22, 2009.

Zeitoun, Mark. (2005). *Hydro-Hegemony Theory – a Framework for Analysis of Water-related Conflicts.* King's College – London.

Zeitoun, Mark (Senior Lecturer in Development Studies, Centre for Environmental Policy and Governance, Department of Geography & Environment London School of Economics and Political Science), personal phone interview by author, Austin, TX, November 26, 2008.

Index

A. Abukhater, *Palestine - Peace by Piece*,
https://doi.org/10.1007/978-3-030-03985-1

MIX
Papier aus verantwortungsvollen Quellen
Paper from responsible sources
FSC® C105338

If you have any concerns about our products,
you can contact us on
ProductSafety@springernature.com

In case Publisher is established outside the EU,
the EU authorized representative is:
Springer Nature Customer Service Center GmbH
Europaplatz 3, 69115 Heidelberg, Germany

Printed by Libri Plureos GmbH
in Hamburg, Germany